Pigment Power

T. Owens Moore, Ph.D.

Zamani Press
P.O. Box 667
Redan, GA 30074
www.zamanipress.com

PUBLISHED BY

Zamani Press
P.O. Box 667
Redan, GA 30074
www.zamanipress.com

Printed in the United State of America

Cover Design by Keith O. Ryales

Library of Congress Control Number: 2020901024

Moore, T. Owens
    Pigment Power
      1. Melanin Research
      2. Politics
      3. Black Studies
      4. Neuroscience
      5. Ancient History
      6. Astrophysics

ISBN 978-1-884897-04-7 (pbk.)

In Loving Memory of

My Father - Joseph E. Moore, Jr.
November 20, 1928 – April 17, 2018

May your spirit continue to provide
guidance from the spiritual
realm of existence.

Contents

ACKNOWLEDGEMENTS

Family is first and they have provided me with the power to possess this pigment. My mother, Queen Esther Moore, birthed me and your continued existence on this planet during this challenging social time is a blessing. Thank you for being the #1 fan. Like the The Intruders sang, I'll always love my Mama. And a special thanks to Sheryl Askew for being the supportive mate in my life and the #2 fan.

All three of my children (Tyehimba, Jabari and Jemila) are deserving of acknowledgement because as you walk righteously, you inspire me to keep doing better. I love each of you and I thank you for being great models for your generation. On both the Moore and the Lee side of the family tree, the roots go deep, and the branches extend wide with the uncles and aunts and cousins who have shown me tremendous support over the years.

In life, we acknowledge those who have provided support for our continued success and longevity. For me, I am thankful for the continual support from key members of the conscious community, partnerships with Zamani Press, dear friends, and members of my Clark Atlanta University family. I cannot mention all of the names, but you know I appreciate our association over the years.

For Zamani Press, it would not be in existence without the powerful vision of Dr. Jelani Madaraka. In addition, the professional skills of Keith Ryales and your creative genius are appreciated. Keith, the innovative designs made by you are always exemplary, and it has been a pleasure to work with you from the beginning to the end. Also, I would like to thank Dr. Vikki Armstrong for her eagle-editorial eye and assistance to complete this project for me and Zamani Press.

I would not be where I am without the encouragement and overwhelming support from the conscious community locally, nationally and internationally. Nia Damali from Medu Bookstore has shown unwavering support to many authors, and I am grateful for her to own the #1 bookstore for me. Along with Medu Bookstore, WRFG 89.3 FM hosts have been strong voices with their stimulating radio programming. I am forever grateful for air shifters like Akhenaten Amen Ra S'L'M-Bey,

Wanique Shabazz, Rah Karim-Kincey, Ahzjah Netjer, Tawhiyda Tupak-el, Karen Marie Mason, D.J. Naysir and Ras Kofi who have provided a platform to support my work.

Even nationally, Lynn Hampton and Vince Robinson in Cleveland and Orisade Awodola in Seattle have been instrumental in allowing me to reach the community. The 360 degree of Truth crew and the I-ACT posse have been respectful to me, and I am thankful for your desire to disseminate my literature. Internationally, the students, faculty and administrators at the Steve Biko Institute in Salvador, Bahia Brazil will always be with my heart from a long distance away.

I greatly appreciate the exposure and it has allowed me to influence new authors on the block who have powerful information to share with the public. For example, I would like to give a pigment power shout to Sabrina Lawton, J.A. Hawthorne, Tamika Moseley, Brenda Tillman and Stephannie Smith. I thank you for believing in my message and keep impacting the world.

I would like to thank my fellow independent scholars, publishers and authors like Mwalimu and Yaa Baruti with Akoben House and Dr. Jeff Menzise with Mind on the Matter Publishing, and Dr. Jewel Pookrum with J.E.W.E.L. University of Immortal Sciences for demonstrating how to produce quality work. Also, the stimulating intellectual discussions with Bruce Bynum, Ann Brown, Chike Akua, Niyana Rasayon, Lennell Dade, April Wells, Asante Bradford, Earle Mitchell and Bill Moore have been useful to help me ground my ideas.

In addition, I am very thankful for the Department of Psychology faculty and students and the CAU administration for providing me with a comfortable intellectual environment to produce this book. The names are many, but I would especially like to thank Dr. Dorcas Bowles for showing financial and moral support with numerous academic endeavors, and for Dr. Medha Talpade for using my literary work for students to read in class.

Since a healthy body is also important for a good life, I would like to thank my multi-connected- tennis contacts in the North and the South for giving me great exercise on the court to stimulate the neurons needed to produce this provocative work. For those who have provided moral support, I apologize for missing any names, but please know you are not forgotten.

Lastly, I would like to thank the living legends and the ultimate Jegnas such as Drs. Naim Akbar and Wade Nobles for extended confidence in this project. Beyond those other scholars who are alive, the souls of Dr. Richard King, Dr. Frances Cress Welsing and Llaila Afrika were present when this book was formulated, but each of these impactful scholars have gone on to the ancestral world. I acknowledge their positive influence on my thoughts.

PREFACE

This book was written and rewritten and put aside and picked back up over several years. I have always wanted to update and revise my other books on the topic of melanin, but time never seemed to cooperate. To complete my next project, I decided to work on Pigment Power to add new information to the discourse.

The information compiled in this book goes beyond what I previously wrote in The Science of Melanin: Dispelling the Myths, The Science of Melanin (The Second Edition), Dark Matters Dark Secrets and Why Darkness Matters: The Power of Melanin in the Brain. I have given numerous presentations nationally and internationally on the topic, and I have the lead article in the African American Wellness Hub called The Biological Psychology of Blackness. The current book has updated information on research that has been conducted throughout the world.

With the tumultuous and changing dynamics in world politics, the astonishing scientific investigations and the horrific pandemic we are experiencing in 2020, Pigment Power is written as a comprehensive book to understand both political issues associated with "blackness" and the physiological advantages of having melanin strategically located in and on the human body. The book is written in a provocative manner to make you think, especially during this unprecedented COVID-19 pandemic.

Actually, Pigment Power was complete and ready to be published before the COVID-19 pandemic hit the world. As the devastating news continued to unfold, I felt it was important to address the topic because it increasingly felt like it was a "Plandemic." Therefore, the last chapter was added to ensure people are knowledgeable and capable of taking care of their health. Pigment Power is all about taking control of your life.

Pigment
Power

INTRODUCTION

Whoever controls the narrative controls the discourse. Therefore, the narrative must change for melanin-dominant people to have a different form of consciousness to navigate the continual assaults on their humanity. To survive the onslaughts to their character, revolutionaries from the Civil Rights movement in the USA often chanted slogans related to obtaining Black Power. It is no longer popular to say Black Power because it has become dated with time since the 1960s. However, we can say *Pigment Power* because it gets to the real nature of the beauty of blackness.

Pigment has many connotations. It is not only a descriptor of "black" or "blackness" because there are other hues beyond black. Primarily, topics on pigment can be highlighted in areas such as health, politics, social studies, economics, religion, education and simply, respect. When the words Black Power were uttered during the tumultuous time period of the 1960s, it was automatically equated with dark-skinned people (i.e., formerly Negroes, Coloreds, Afro-Americans, Black Americans, African Americans) responding to the oppressive conditions created by melanin-recessive people. The homegrown as well as the globalized terrorism of white colonizers was met with a conscious awakening that black is not bad. Black Power is still needed, but from our science perspective, *Pigment Power* can mean the same thing in this era.

For example, living in America after the historical eight-year term of President Barack Obama, it has become increasingly clear on how important it is for melanin-dominant people to be unified against the constant onslaughts to their existence. Pigment politics are real, and the volcanic eruption of hatred in American society after the first pigmented family vacated the White House is despicable. The lava and the ash have wreaked havoc on the core of our humanity. The use of the title *Pigment Power* has the potential to explain the problem and to neutralize the threat of this volcanic disaster that is trying to destroy the existence of pigmented populations on the planet.

This book will combine a discussion on both the social and the health consequences of pigment from every aspect of our experiences in life. When we say health, we are referencing how a lack of pigment in the physiological system can cause negative consequences. In the social arena, the outside imagery of skin color has a strong impact on politics and identity. *Pigment Power* is not to express superiority, but it is a profound expression of the importance of melanin in overall health promotion.

Why are our hands two-toned? Why are the soles of our feet two-toned? Why does the male penis have a multicolored appearance? Why do the nipples and the vaginal area of females have a different toned appearance? The external areas of the body are sensitive to extrasensory perception. Feet to the ground, hands to the air and the penis in the vagina are spiritually dwelling experiences that connect us to nature and procreation. The reality is that our bodies are in tune with cosmic forces, and the skin is a physical conduit possessing melanin that enhances the harnessing of the energy that pervades the Universe.

In *Pigment Power*, we will discuss the science and health consequences of having melanin in twelve chapters. On a world-wide level, the fragility of xenophobic melanin-recessive people has helped to construct a very real caste system that subjugates people with a dark hue on the lowest realms of the socioeconomic and political ladder. Therefore, we begin with a sociopolitical analysis in Chapter 1 - The Social Implications of Blackness.

Chapter 2 - The Role of Melanin in Health Promotion highlights research articles on how the presence of the pigment melanin is critical for life and overall health. This chapter is critical for understanding the thrust and science behind the multiple topics that follow on pigment power.

Chapter 3 – Melanin and Nervous System Development presents information on the brain and how any disruption in melanin formation in the development of the brain can impact nervous system functioning. Also, we do not think about the importance of light, but there are factors that affect our mood and awareness, so Chapter 4 - Sunconscious, will delve into the effects of light on life. Considering this importance, Chapter 5 – Electromagnetic Energy and Mental Health will go into more depth with other forms of energy which also impact life. Adding to this topic of energy, we further explore how psychomotor skills are heightened in Chapter 6 – Rhythm Nation.

In Chapter 7- Penis Power and Pigmentation, we present a provocative discussion on the role of a melanin-dominant physiological system and the societal implications of male dominance. In Chapter 8 – Melanin and Cellular Enhancement, the health benefits of melanin on a cellular level will be presented. In addition, the effects of stress on the human experience will be presented in Chapter 9 – Melanin Protection, Stress and Aging.

As we move out of the body into technology and cosmology, the last three chapters provide a different discussion on pigment power. Graphene is a black, carbon-based material in science and the natural energy of melanin is harnessed into material science. Chapter 10 – Technology and the Blackest of Black, will present research on graphene as a carbon-based game changer in science and highlight specific science that is based on what has been stated for years about the science of melanin. In Chapter 11 – Astrobiology and Melanin, there will be a provocative view presented on the order of our Universe. We will discuss alternative views on how creation was manifested in the cosmic order, and how less popular views on creation impact our revised understanding of human development. In Chapter 12 – Pigment Powered Defense System, we will discuss what is really going with the unprecedented coronavirus pandemic.

In sum, welcome back for another "Endarkening Experience."

T. Owens Moore, Ph.D.
Redan, GA

CHAPTER 1

THE SOCIAL IMPLICATIONS OF BLACKNESS: FROM THE CELL BODY TO THE BODY POLITIC

Be who you are or be controlled from afar.

Part I – Pigment Politics

Around the world and in many regions of the Earth, there is a strong visceral reaction to skin color. It is unfortunate that skin color has been erroneously connected to and often correlated with thinking capacity, but this is the case. It is ludicrous to believe skin color should be such a strong factor affecting social relationships, but it has developed into a serious human relations dilemma. It is like a nasty virus that keeps reappearing. Historically, it has not always been the case where people despised black-brown-dark skin color (Kelly, 2011). Somewhere along the way white people successfully and collectively organized the pervading thought that black-brown-dark skin was bad (Welsing, 1991). In Part I of this chapter, there will be a brief explanation of the problem. Blackness has been equated with something bad and this belief must be dismantled and neutralized. In Part II, there will be discourse on contemporary political events (e.g., the Obama Presidency) which changed the course of human history. In Part III, a comparison and contrast of the simple explanation using the microcosmic cell and the macrocosmic world of politics will be presented. Retrospectively, a paradigm shift has occurred, and it is time to move towards a new view of humanity.

From a geopolitical perspective, the construction of a white supremacist society has the majority of the world's population believing that pigmentation is critical to behavioral functioning. In this context, white supremacy can be used interchangeably with the term racism. It is a very powerful concept and it is quite spectacular for the entire world to be tainted by the propagandistic culture of white supremacy. The corporate controlled media has the power or ability to define reality and this is a major way that false images can be disseminated to manipulate

7

the masses. The antithesis of white is black, so psychologically, it is simple to process how the darkest people on the planet are denigrated in a white supremacist cultural orientation. Reverse psychology is the only way to change a view that black is bad. Historically, when white Western European scholars from Greece began controlling the dissemination of knowledge in ancient history, black and anything dark was associated with something negative and bad.

Once people believe that black-brown-dark skin is bad, a host of retarded, backward and illogical concepts are developed. The most obvious fallacious argument is the concept of race. Race is a sociological construct that has no basis in science. Race was created to meet the needs of a European thought pattern (Ani, 1994) that required entities to be split and divided. Views and perspectives on life will always be in fluenced by how the mind was initially programmed.

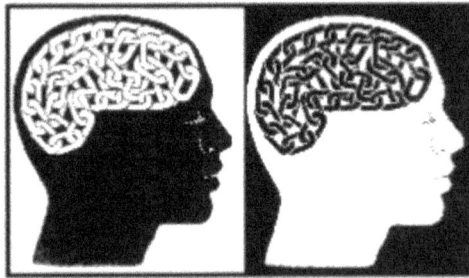

Fig. 1 - Psychological Chains.

Programming is a continual process; however, it is the past programming that sets the stage for how every subsequent and future program will operate. When people are miseducated to associate black with bad, the mental stage is set for dysfunctional thinking. The outcome is a racist society.

To reverse something is to go in the opposite direction. People can be controlled psychologically because it is easy to manipulate the mind and subsequently, the thought process. To be continually told that black is bad, it is easy to generate a reverse psychology to work for the advantage of white supremacist thinking. No matter a person's color, people can collectively believe that black is bad. You can be a liberal or a democrat, a conservative or a republican and still believe black is bad. You can live in any region of the world and believe black is bad.

8

How could this be? Corporate control and the global impact of white supremacy (Fuller, 1969) along with the use of reverse psychology are the answer. The reality is "blackness" is a biological adaptation. It is actually good and without pigmentation, life as we know it would cease to exist.

The blackness we see on the outside of the body derives from the skin. The skin is a magnificent organ that contains many chemicals that also exist in the brain/nervous system. Inside the skin is a complex matrix of cells that keep our body intact and flowing with life. Deep within the skin are cells which produce melanin and they are called melanocytes. These intricate cells look similar to brain cells (i.e., neurons), and melanocytes secrete melanin to create coloration. Melanin, therefore, provides the body with the physical representation of blackness. Beyond the skin, "What provides a person with the cultural experience of blackness?" This is a question that cannot be clearly delineated in these short pages because there are a host of experiences over a lifetime that influence the cultural expression of blackness. It is the multifaceted experiences that lead to the social implications of blackness. If we were to subtract the negative views about black or blackness, then this would be the beginning stages of developing a new paradigm shift in consciousness.

In this section, the emphasis is placed on life because the pigment melanin is a complex biopolymer that is critical for the maintenance of life. I have researched the topic of melanin for nearly 20 years (Moore 1995; 2002; 2004; Bynum, Brown, King and Moore, 2005) and there is an abundance of research to study. However, this chapter is not written as a review of the literature. As I have written for years, melanin has a significant role in the early stages of fetal development, and there are biological advantages to a properly functioning system with melanin. Any loss of melanin internally or externally can cause abnormal physiological functioning.

From a variety of scholarly perspectives, many people are being guided along the color line to explore the experience of blackness. The theme of this text is specifically related to blackness and the topic in this chapter is skin pigmentation. Pigmentation is critical for the physiological functions of the human body, but we should not equate melanin with the reified term called intelligence. Melanin in the nervous system, however, can play a critical role in psychomotor abilities and should never be considered a waste product in the body. The next section will define a polemical issue in our current society to shed light on the social implications of blackness.

Part II – Social Politics

James Brown popularized the slogan, "Say it loud I'm black and I'm proud." What followed were prideful comments like "Black and Beautiful." These slogans were not reverse racism; they were verbal reactions to a white supremacist society that systematically denied blackness. Because there are social implications related to being black-brown-dark in color, we have to ask the question, "What do these slogans really represent?" The attempt to highlight this rhetoric in this chapter is to demonstrate the razor sharp, sensitive feeling that arises with any discussion on blackness. What emotions are generated when white people utter the words "White and Beautiful?" If there were constant chants back and forth expressing one's ethnic pride, would it be a simple knee jerk reaction to the opposing slogan, or would it be a supremacist type of thinking that goes into bullying and projecting personal pride upon someone else? Since racism or white supremacy is operational (Welsing, 1991), we need to understand how people express their pride. What does it really mean to express one's blackness and say, "Black and Beautiful?" What does it mean for a white person to express their whiteness and say, "White and Beautiful?" Either way, both views could be the direct result of a racist, white supremacist world that has been created. Similar to its creation, it can also be dismantled.

Historically, the expression of blackness has been an effective way to repel racist attacks and to simultaneously remove a person from the mental shackles of oppression. When expressing blackness, is it an experience of pride or is it a mental reality surging from the unconscious mind? The blackness described in this chapter is referring to the pigment called melanin that exists in the human body and the culture that goes along with being a person with various hues ranging from light-brown-black- or dark-skin. Pigmentation from an African ancestry is the key and that is why there are variations in appearance to define blackness. The hue or pigmentation relative to outside appearance is

10

what has become a critical factor defining social relationships. We must logically process how this skin color gradient has affected the way humans communicate and relate to each other. Let us look at the political situation in the United States of America (USA) with the presidential transitions from the last four administrations.

In a nutshell, we have seen the full display of pigment politics from former Presidents, Bill Clinton, George W. Bush and Barack Obama and now Donald Trump. Under each president, there was a paradigm shift and a vexing change in the public opinion concerning issues related to people of color living in a white-controlled society. It is interesting to note the flip-flop change from Democrat (Clinton) to Republican (Bush) to Democrat (Obama) to Republican (Trump). It is also interesting to note two of these past presidents were impeached (Clinton and Trump). Although there has been a waxing and waning of political positions, there has been the consistency of the white suprem-acy virus popping up via obverse political decisions toward melanin-dominant people, the continual neglect of the poor, racially-motivated mass shootings, the invasion of black nations and the overall negation or lack of support for reparations for the ancestry of this nation's formerly oppressed people.

Under the Bill Clinton administration (1993-2001), there was the curse of mass incarceration policies that came back to haunt former First Lady and Secretary of State Hillary Clinton in her unsuccessful bid to be president in 2016. Since Hillary Clinton was on record for labeling some black people as "Super Predators" that needed to be contained and locked-up, and these policies under her husband's admin-istration decades earlier ruined the lives of many black families. Under the George W. Bush administration (2001-2009), the negligent response to some Hurricane Katrina victims left a stain on his presidency where citizens and celebrities publicly declared he did not care about black people. This was strange, however, because he had two very prominent and respected African Americans, General Colin Powell and Dr. Con-doleezza Rice to serve, in his cabinet as the Secretary of State as well as National Security Advisor. Despite their presence, the devastation left behind in the black communities from Hurricane Katrina in New Or-leans, LA was deeply entrenched in pigment politics.

Transitioning to the administration of Barack Obama, erroneously coined the first African American Commander-in-Chief (2009-2017), there were great expectations for significant change in US politics. What we experienced was a different level of pigment politics because he was the only one of the last four presidents that actually had pigment in his skin. He was half white and half black, so the excitement alone was expected to eliminate the white supremacy virus. In contrast, Ta-Nehisi Coates (2017) refers to the eight years of Obama's presidency as an "American tragedy." As we relate this topic to pigment politics, Donald Trump becomes the white backlash to a "black man" who was previously in power. Under Trump, the most virulent form of the white supremacy virus was unleashed to both the nation and world. To explain this selection of an unqualified candidate who was eventually impeached, we must go back to the global experience under Obama.

It is significant to analyze the current paradigm shift in the political scene in the U.S.A. from Clinton to Bush to Obama to Trump. The election of a person of color to be the President of the USA was witnessed by the entire world and the selection of Barack Obama was diametrically opposed to everything supporting a white power structure in politics. The election of a nonwhite president to lead the nation from the White House greatly complicate the social implications of blackness. White is a color also, so who was really in the White House? Was it a person who displayed blackness or was he just a person with black-brown-dark skin with no identity? The view must change because the old paradigm no longer applies that black-brown-dark skin equates to unintelligent and down-trodden people. The paradigm shift that was occurring would require in depth research on the consciousness of people worldwide as well as an inquiry to determine what type of shift was actually occurring.

I was 44 years old during Obama's first Inauguration. I was asked, "How does it feel to see a 'Black Man' as the 44th leader of the USA?" In one sentence, I would say, "I am cautiously optimistic and not overly ecstatic." Obama was the best candidate for change in the 2008 General Election. I read Obama's second book (2006) from front to cover, and he was overwhelmingly the best candidate for me to select. There was absolutely no hesitancy to select Obama to serve as the 44th

President of the USA. As a conscious person of African descent, however, I was fully aware that he had taken the helm to be the leader of the greatest purveyor of violence on planet earth. He would have a daunting task ahead, and it would be best to provide him with support as he attempted to stimulate change.

In terms of the social implications of blackness, the occasion was both historic and symbolic for me. In the year that I turned 44, the selection of the 44th President of the USA was a person who resembled me. However, we should be clear about his ethnicity. The confusion is around the term African American. According to the definition of who I am, he does not have the same ethnicity to be called African American. Obama's mother is a White American and his father is a native African from the country of Kenya. In contrast, my mother is African American, and my father is African American. I am black-brown-dark in color like Obama, so does that officially make Obama African American? If that is the case, white colored people who migrate to the USA from Angola, Mozambique or any part of Africa would be considered African American just like me. Yes, very confusing as we discuss the social implications of blackness. When Obama was made to publicly denounce the minister of his church (i.e., Rev. Jeremiah Wright), the blackness he obtained from his spiritual experience at Trinity United Methodist Church began to get white-washed away. It was clear it was a political decision to denounce his former minister, but it left a searing mental stamp on the social implications of blackness.

Concisely, Obama's mixed parentage does not automatically equate his lineage with the normal description of African American. Therefore, the media is incorrect to call him the first African American President. The work of the prolific writer Joel A. Rogers (1965) and his short pamphlet called *The Five Negro Presidents* sheds light on a history that has been neglected by the media. In fact, the historical legacy of John Hanson may be more accurate for those who understand the founding of the USA as confederated states. Hanson was considered to be a man with pigment while serving under the Articles of the Confederation, and he was elected to serve as the first President prior to George Washington. Actually, there were eight other presidents before

George Washington. To conclude my remarks on Obama, his presidency added to the complexity of the social implications of blackness.

Considering the paradigm shift where blackness can no longer be equated as inferior, the thinking must change. The view of what is called blackness is different for white and nonwhite people. As a matter of fact, there are a host of permutations that exist. For example, people of color who have been overwhelmed by white supremacy may hate their colored body and succumb to self-hate. People of color who speak with pride may love their colored body and rhetorically say, "Black and Beautiful." White people who have been indoctrinated to believe black-brown-dark skin is bad may become more entrenched into a greater xenophobic perspective. On the other hand, prudent thinking white people may realize they are not the majority of people on the planet and willingly accept the equality that exists in the human spirit (Wise, 2011). No matter how you calculate these complex scenarios, change is coming. No predictions are made in this chapter, but paradigm shifts have occurred from Clinton to Bush to Obama to Trump.

As Welsing (1991) has definitively claimed for years, white people have responded with a psychological feeling of inadequacy about their appearance. With their low birth rate and lack of ability to produce melanin, white people have collectively and psychologically constructed a world in which they can remain in control of resources, and concomitantly, depopulate the world of people of color. By claiming nonwhite people to be inferior, it makes it easier to label black-brown-dark skin as the abnormality. The truth of the matter is that having pigmentation is a necessity for LIFE.

Part III – A Paradigm Shift on Blackness

To have a better quality of life for people all over the world, a shift to appreciate the beauty of blackness is needed. In fact, during a world-wide climate of attacks on melanin-dominant people, history was made in the selection of the 2019 Miss Universe. From South Africa, Zozibini Tunzi was the dark-chocolate contestant with "ALL natural" looks and a campaign platform focused on reducing racism. It is impossible to appreciate the branches of a tree without first loving the roots of the tree. For so long, black-brown-dark-skinned people have

14

been taught by white people that the color of a person's skin makes a difference in life. Therefore, to have a good life, everyone wants to be associated with what is believed to be better.

Many white people have been duped by their own theories, and they have convinced many people throughout the world that anything associated with brown-black-dark skin is a problem. Rather than seek unity and build a better nation and world, many white people in the USA sought White Power organizations and bought weapons because they had an irrational fear about a person of color in the White House. This xenophobic reality is tantamount to a mental illness, and this paranoia makes sense when the society is socially sick and miseducated. Since we live in perilous times, wise thinking is needed to bring unity to this planet. Obama's multiethnic background was the best example of what happens when we put color aside, but the white supremacy virus still flourished. We must understand that he served the USA and not Black America, so Obama should have been the perfect leader to mentally usher in change for a better quality of life for people throughout the world.

It is obvious that if Obama did any of the awful shenanigans that Trump did prior to and during his presidency that Obama and his blackness would have been vilified. Trump was elected on a white privilege pass, and he was not unanimously impeached due to a white supremacy pledge. History will record that Trump tried to reverse the social progress made by his predecessor and ushered in a bad contagion that has contaminated world politics and devastated international relations. If Obama openly expressed grabbing women by their private parts and then was found guilty of impeachable offenses, then we would be back to the Reconstruction Era when the white establishment claimed that black leadership would ruin the USA. In reality, Obama displayed respectable leadership, but, the question remains "Was he able to fully express his blackness?"

There are negative consequences related to avoiding the natural expression of blackness. For example, avoiding your culture, running from your reflection, bleaching your skin, altering your facial features and mentally identifying with a reality that is not yours are definite ways to lose your blackness (Akbar, 1984). As a real possibility, imagine if

15

Obama's wife was melanin-recessive or white like his mother and he openly denied his blackness. The odds are his image would be totally different to the world. Throughout his lifetime, there was always that possibility that he could lose his blackness, but there were some anchors in his life that did not allow him to float away (Obama, 2006). Michelle, his wife, is 100% African American (Obama, 2019) and the church he chose to ex-communicated himself from in Chicago was truly a part of the African American community. Therefore, there is no need to debate if he is black or not. Under his leadership, the question will be, "Did the quality of life for all human beings, starting with those who have historically remained on the bottom since the creation, improve?" Progress should be for everyone and not some. When a paradigm shift is on the horizon, this does not mean white people need to be fearful. Although some white people may revert to paranoia, it is time to educate people on a world-wide level to understand the beauty of blackness.

White supremacy or racism is the main reason why people who exhibit blackness are typically at the bottom of any society you visit in the world. Over the million plus years of human development, this has not always been the case (Walker, 2011). A correct reading of history will reveal that white supremacy has been a masterfully administered technique to reverse the psychology of people no matter what color they exhibit. Everyone wants a good life, and the good life can be sought socially, economically, politically, spiritually and physically. To reiterate, white supremacy or racism is the reason why black-brown-dark-skinned people are always at the bottom of these categories. It may seem like an oxymoron, but it is the blackness that has helped people to survive. There is beauty in blackness, and if you appreciate the purpose of color in life, then it is easy to see the power of pigmentation

Globally, there is no denying that the entire world is infatuated with blackness (even if it is unconscious) because we see Hip-Hop culture has penetrated every corridor of the globe. Even though the world relegates black-brown-dark-skinned people to the bottom rung of society, it is the creativity or blackness that people desire to emulate. From the cell body to the body politic, we want to visualize how ingenious the cells of our body work and how ignoramus people act in politics. A properly functioning system of equilibrium can only exist when people or entities work toward obtaining balance. Concisely, the

Obama Presidency equated to balance, and the Trump Presidency equated to chaos.

Balance is critical, and as an example, the cells of the human body create homeostasis without direct awareness. The cells of the body know what to do in order to survive. It is the host and its aberrant behavior that alters the natural order. Similarly, racism or white supremacy has altered the natural order of life and it must be neutralized to usher in a new paradigm shift in human relations. I am not naïve to believe we can completely eradicate racism or white supremacy. Racism can appear like a virus which can cause a disease called *Wetiko* (Forbes, 1992). Jack Forbes, the native American intellectual, calls *Wetiko* the diseases of aggression against other living things, and more precisely, the disease of the consuming of other creatures' lives and possessions. Whatever we call it, this disease, this *wetiko* psychosis, according to Forbes, is the greatest epidemic of sickness known to man. Since viruses can attribute to epidemics, and a virus can arise at any moment if it is not contained, neutralization is critical. In sum, we can learn from the cell body to affect change in the body politic.

From the Cell Body to the Body Politic:
Mental Liberation is that Simple

Life is totally dependent upon a well functioning body and there are internal bodily mechanisms that transpire everyday, every second and every moment to keep the body alive. The internal mechanisms of the body are functioning without your awareness. In my discipline as a neuroscientist, the internal mechanisms are extremely intricate and detailed at the cellular and molecular level. Without these internal mechanisms, there is no body and no life, and this is tantamount to the experience known as death. As a trained neuroscientist who is interested in African-centered issues, I am forced to think of these two very different areas in my daily thoughts. I am trained to study the neural mechanisms of behavior, but the internal dynamics of my consciousness leads me to focus on African-centered issues that will potentially lead to the liberation of oppressed minds. From the cell body to the body politic, we can use this analogy to solve many of the problems we experience in the world.

How does a cell work and operate? How does the world function and operate? Is the word politics appropriate for the way a cell operates? These are questions to lead us close to a better way of getting the global community to operate more effectively. I admit that using the phrase from the cell body to the body politic is a play on words. However, politics is the operative word. Politics usually pertains to issues that are impacting or affecting the normal flow or natural order of social relations. The dictionary definition of politics pertains to having practical wisdom or being prudent, shrewd, and/or diplomatic. Humans have been on this planet for thousands of years and never in the history of man have we had such gross distortions on what it means to be a human. There are too many political issues facing humans and confusion is around every corner of our existence. The corporate control of media has many of us consuming a diet of confusing images of what is right and wrong. What we consume physically, mentally and spiritually can deter us from reclaiming and revitalizing our communities when the source of our diet is not under our control.

Cells in the body can work together to benefit the whole. In a millisecond portion of time, for example, a big corpuscle can patiently wait to go through a tight artery before the smaller corpuscle travels through the bloodstream. On the surface of the body, an epithelial cell can slough off the surface of your skin in the form of ash to make space for the new cell to provide beauty in the external appearance. When cells in muscle sense a dehydrated environment, neural impulses are sent to specialized areas of the brain to initiate the thirst drive to unconsciously seek water for survival. There are a host of other cellular examples to provide, but the point is that various cells in the human body are working together to keep the body functioning properly. What type of intelligence could be in these single cells that humans seemingly cannot manifest in the social world? On a global scale, humans are killing one another and destroying the environment. Why does it appear as though the large human brain cannot muster the intelligence that single cells have? As summarized in the chapter, POLITICS is the reason we cannot seem to generate solutions to many problems in society.

Personally, I cannot save humanity, but I can leave this world better than it was before I arrived. The issue is in the tissue and there resides the problem. Our tissues are bombarded with a poor diet and we MUST make changes internally if our lives are to change. Focusing on internal reparations is the reminder that the media can be responsible for what we are not, but we are individually responsible for what we ought to be. Let's stop playing politics with our future and focus on restoring as well as reclaiming and revitalizing our blackness and our intelligent human spirit.

In conclusion, the cells in our body provide a perfect example of how the body politic can operate intelligently. The cell body knows how to intelligently eliminate foreign intruders and biochemical agents that should not be in the body. Our internal cells revitalize and reclaim a positive environment on a daily basis. For some reason, our much larger brains have difficultly finding intelligent responses. Corporate control of the media is a foreign agent that has infiltrated the minds of many people. We must develop the same intelligence of our individual cells to seek life. It sounds easy, but again, we allow politics to control the natural intelligence of how humans should function together to benefit the whole.

Just as Hip-Hop has changed the world, we as people of African descent will continue to change the world. Note that cosmic shifts on a planetary level can happen over millions or hundreds of years whereas paradigm shifts in social consciousness can occur in a matter months, days and even moments. We all saw prior to his selection to be the 44th President of the USA, Obama toured other countries, displayed his regal blackness and gave people throughout the world hope and confidence that his blackness was reassuring and not repulsive. By reusing the "Southern Strategy" to get elected, Donald Trump, also known as #45, ushered in a paradigm of hate and ignorance and, unfortunately, other countries throughout the world have shifted in this direction. From the cell body to the body politic, let us eliminate the virus of hate and ignorance which causes the Wetiko disease and find the beauty in blackness to build a new humanity.

CHAPTER 2

THE ROLE OF MELANIN IN
HEALTH PROMOTION

The power of pigment is penetrating.

As a complex biopolymer, melanin has a major role to play in supporting life. Melanin-dominant individuals can subjectively experience the health promoting effects of melanin whereas melanin-recessive individuals can only objectively speculate on how to control and manipulate the experience. Over decades of scientific exploration, melanin-recessive scientists have had minimal interest in melanin or any studies dealing with blackness. Therefore, there has been resistance to acknowledge the importance of melanin in the general literature.

In the ethnocentric explorations, African-centered melanin-dominant scholars have been at the forefront of promoting the health benefits of melanin (Barnes, 1988; King, 1991, 1994; Pookrum, 1993; Afrika, 1989, 2000, 2009; Moore, 1995, 2002, 2004; Bynum, 1999, 2005, 2012; Meningall, 2008; Diouck, 2018). Therefore, we have a plethora of authors who have documented their perception of melanin as a life-generating molecule. This author has emphasized the three critical roles of melanin functioning (Moore, 1995). First, it can serve as a neutralizer of toxic substances. Secondly, it can serve as a nerve conduction facilitator. Thirdly, it can function as an energy transformer. For the promotion of health, it must be clearly understood that life would cease to exist without pigmentation.

A major problem for melanin-dominant people living in a white-oriented world is the foreign diet. The food we eat should nourish an internally pigmented system, but many melanin-dominant individuals eat and consume death-enhancing versus life-producing foods. The quality of life can be enhanced when we eat live foods from the earth and not from a box or a can. Moreover, raw foods are critical to feed and nourish the natural beauty of melanin (Diouck, 2018).

If we know pigmented foods can augment positive health (Joseph, Nadeau and Underwood, 2002), we need to emphasize the consumption of pigmented food items such as fruits, vegetables and natural items like mushrooms. The melanin diet has been suggested by others (Meningall, 2008) and the negative effects of a poor diet have been documented (Afrika, 2000). Melanin absorbs toxic substances, and the more toxicity absorbed by melinated cells is a death sentence for the cellular components of the body. For overall good health, decreasing items like dead meat and artificial substances is desired (Joseph, 1989). Whitaker and Fleming's (2005) book called *MediSin* clearly describes the dilemma in the healthcare industry and how the enhancement of life is not beneficial for a capitalistic system. The industry cannot make money unless you are sick and unhealthy. The next five sections will describe a positive way to promote health. A health problem will be used to describe the diseased-state or medical phenomenon, but the description will highlight the positive role of melanin in health promotion.

Rickets (Solar Generation)

Rickets is a bone deformity resulting from a lack of sunlight. Therefore, inadequate solar power to the human body can have devastating consequences and create a host of biological problems. For example, a lack of calcium and vitamin D can lead to rickets during normal development and osteoporosis in older people. Essentially, brittle bones can lead to poor health and diet is a way to offset a lack of sunlight in the human experience. Vitamin D is not found naturally in many food items, but it can be fortified in milk substances, yogurt, orange juice and cereals to comingle with calcium to avoid brittle bones. Fatty fish and portebello mushrooms are also a source of vitamin D. As emphasized before (Moore, 2002), vitamin D is actually a steroid hormone and the name soltriol has been coined by other researchers (Stumpf, 1988; Stumpf and Privette, 1989). The numerous roles of soltriol cover immune boosting effects, mood enhancement and extrasensory perception (Moore, 2002), so there is a physiological reason why you feel better on a sunny day (Figure 2).

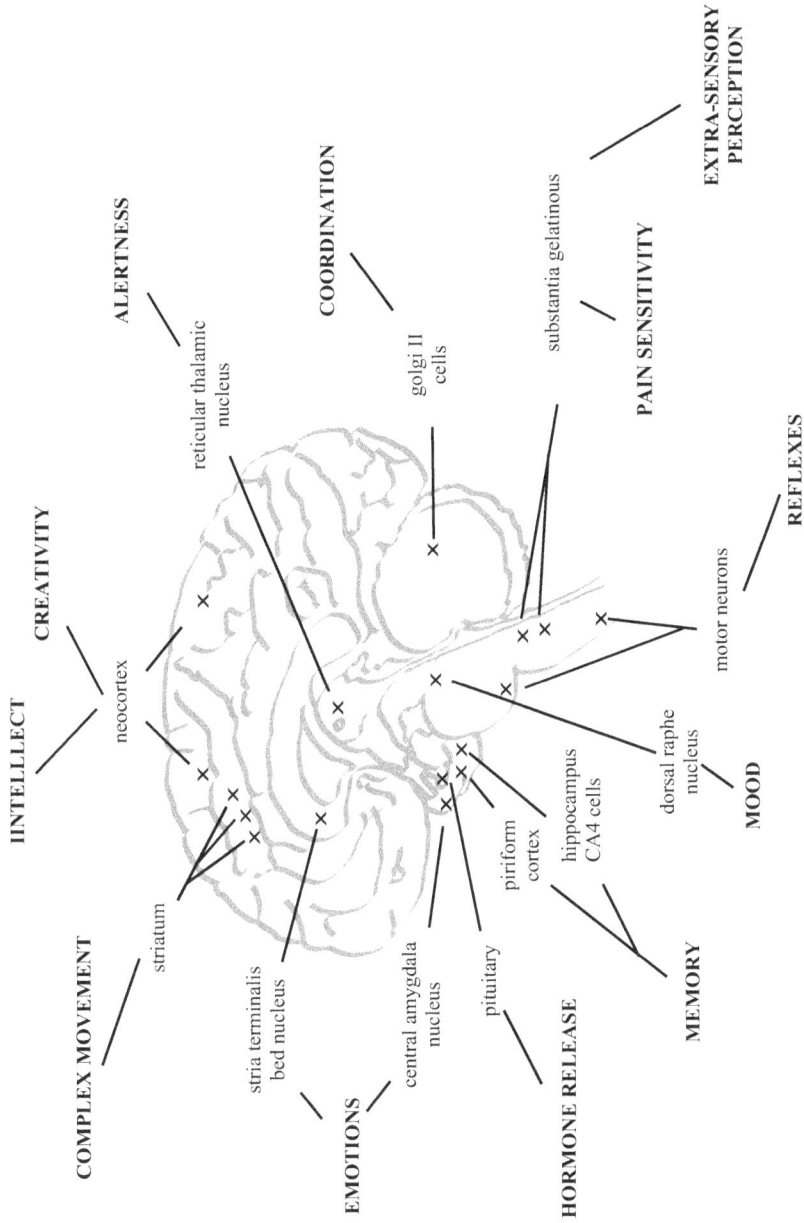

INTELLECT

CREATIVITY

ALERTNESS

COMPLEX MOVEMENT

COORDINATION

EMOTIONS

HORMONE RELEASE

MEMORY

MOOD

PAIN SENSITIVITY

REFLEXES

EXTRA-SENSORY
PERCEPTION

reticular thalamic
nucleus

neocortex

striatum

stria terminalis
bed nucleus

central amygdala
nucleus

pituitary

piriform
cortex

hippocampus
CA4 cells

dorsal raphe
nucleus

motor neurons

golgi II
cells

substantia gelatinous

Fig. 2 - Major brain sites for Soltriol receptors and the correlation with behavioral functioning.

23

Beyond food and the type of diet you consume, daily exposure to the sun is critical for maintaining a functional internal and external pigmentary system. The sun is a physical source of energy, and life would not exist without this solar generator. Melanin in the skin can function as a magnet to absorb the powerful effects of solar energy. The effect is obvious because you feel so much better when your hue is enveloped by the rays of the sun.

In terms of rickets, depending on what hemisphere or side of the earth a person may live, the location can impact the effectiveness of absorbing the proper energy from the sun. We are aware that the sun can produce dangerous ultraviolet light and then cause cellular damage and this type of damage can lead to cancer. Furthermore, an individual may receive inadequate light from the sun, and this can create a different set of problems. As previously stated, the skin generates vitamin D or soltriol from the sun and some positions of the earth lack proper emission of adequate sunlight. In sum, rickets, bone deformities, sad feelings, and a weak immune system are a host of problems that can result from a lack of sunlight. Melanin-dominant people have the sun as a friend and in the spirit of Fela and his classic song Water Get No Enemy, we say here "Sun Get No Enemy."

Neurodegenerative Disorders (Neuroenhanced Experiences)

There is clear evidence that the absence of melinated cells in the brain can lead to debilitating life experiences. It is very easy to focus on the diseased state and see the absence of good health in disorders like Parkinson's Disease. For this section, we do not want to focus on the debilitating nature of neurodegenerative disorders. In contrast, in the same brain areas that are negatively affected, we want to focus on neuroenhanced experiences. Many contemporary melanin-recessive people have looked at anything outside of white as negative and bad. The script has been flipped to make good look bad and bad look good. Once we remove this reverse logic from our brains, we then set the stage for understanding the role of melanin in health promotion.

For example, enhanced psychomotor skill development is found throughout melanin-dominant people from birth out of the womb until death in a casket. People of African descent have changed the world in music, dance and sports. There are not many people on this planet who can deny this fact. How melanin-dominant athletes and performers have changed the capitalistic enterprise is overwhelming. The display of melanin-dominant influences on all forms of music, dance and athletic performances (Rhoden, 2006) has changed the world and enhanced the human experience for all people on the planet.

We can look at the quarterback (QB) position in the NFL that has been traditionally set aside for the white male athlete. No one is denying that white athletes can perform. In fact, Tom Brady, the ex-QB for the New England Patriots, will go down as the greatest of all time because he is the ultimate winner. Additionally, at the college level, the Heisman Trophy Winner for the 2019 season was the unanimous pick for Joe Burrow from LSU. With the other white NFL QBs who have dominated the sport over the history of professional football, we have been thrilled by their performances. However, when the melanin-dominant athlete is given a chance to display their neuroenhanced skills, the entire game changes for the team, for the fans, for the excitement and commercial appeal, and for the money to be made in this capitalistic system.

James Harris, Warren Moon, Randall Cunningham and Doug Williams were prominent examples of black QBs who preceded the current slew of black QBs in the NFL. The list of contemporary and influential black QBs is numerous, but the ones who have been champions and made their teams contenders for the playoffs have been Russell Wilson, Patrick Mahomes, Deshaun Watson, Teddy Bridgewater and Dak Prescott. Prior to this contemporary list, sport fans we were mesmerized by Michael Vick as he blazed by defenders and changed the game, again. The unique skill set to pass, throw, run and intelligently make a decision with the ball has been taken to a new level each time a melanin-dominant athlete is given a chance to compete. Even in 2018, Patrick Mahomes from the Kansas City Chiefs earned the Most Valuable Player (MVP) award for the NFL.

In the case of Michael Vick and his career, he was drafted as the first overall pick and he played for 13 seasons primarily for the Atlanta Falcons. He was prosecuted for dog fighting, served his time in prison, and then returned to the NFL to play for several teams. After the "dog" issue he was never the same player on the field, and he has never recovered from the majority of melanin-recessive dog lovers who seem to continually vilify him. It is mind-boggling because these same people who protest and deny him the opportunity to receive any kind of athletic award can blindly justify the impeachable offenses and misogynistic attitude towards woman displayed by #45. Apparently, pigment power is on display because a black man violating a dog is worse than a white man violating a woman.

Whatever the political controversy, Vick changed the game and fans could not imagine seeing this type of player again, until along came Lamar Jackson, a standout for the Baltimore Ravens. Initially, he was denied the opportunity to try out for the QB position, but he insisted. Jackson has taken the QB position to another level beyond Vick. Due to his dynamic play and record-breaking performances, Jackson earned the 2019 NFL MVP award. During the season, one crazy narrative by a white 49ers broadcaster (i.e., Tim Ryan) was that Jackson was good at the ball fake because his dark skin with a dark football with a dark uniform makes it difficult for the opponent to see the ball. The suggestion was that Jackson's skin color gave him an advantage. Ryan is a component of the white-controlled media, so even though it was a ridiculous statement, having access to the public narrative can allow you to say unconsciously racist statements. The NFL and all sports need black athletes, but if black athlete were to stop participating in a slave-oriented arrangement for the white capitalistic system, the system would collapse.

The former ESPN SportsCenter anchor, Jemele Hill, raised the question in a provocative article in The Atlantic (Hill, 2019). With football and basketball, it would be truly revolutionary for black athletes to STOP and create their own leagues without white capitalistic control. These two aforementioned sports are the two money makers for the slave-oriented capitalistic system. Baseball is the American sport, and although melanin-dominant athletes have excelled in baseball, it would have indeed been revolutionary to keep the Negro Baseball League decades ago. Sports like hockey and golf are white-dominated sports, but the influence of black athletes has influenced both sports. On a global level, look at the power of a pigmented player in golf like Tiger Woods. Woods became the capitalistic face of this sport. When he faded, ratings faded, and the fan base changed. That was certainly not good for capitalism.

In contemporary times, we can go on to tennis (Venus and Serena Williams), track and field (Usain Bolt), gymnastics (Simone Biles), swimming (Simone Manuel) etc. When melanin-dominant people are given an equal nonracist chance to compete, they change the game forever. In the last decade (2010-2019), Serena Williams earned the Associates Press (AP) Female Athlete of the Decade Award and Simone Biles was runner-up. Simone Biles won the AP Female Athlete of the Year in 2019 and Kawhi Leonard won the top male Athlete of the Year. It was Lebron James who won the AP Male Athlete of the Decade. If the white capitalistic enterprise cannot control melanin-dominant athletes, then black athletes becomes useless to the slave system - Ask Colin Kapernick.

Intellectually, the creative genius of melanin-dominant people has been exploited and many inventions were stolen from melanin-dominant people. Even in this century, there is still a battle to assert the mental strength of melanin-dominant people without needing to discuss superiority. If one just looks at the accomplishments of a people that have been historically exploited, imagine what could happen if oppression was not part of the equation? Melanin-dominant people would build pyramids again, study the cosmos and teach other groups of people how to be civilized. Historically, other scholars (Bradley,1978; Barnes, 1988; Welsing, 1991) have written why some groups of people who lack melanin have less civil ways of dealing with the human experience, so pigment power can be neuroenhancing on multiple levels.

Hypertension (Tension at Ease)

The vessels or channels in our bodies carry the life fluid called blood to every portion of the body. Without blood circulating, life would cease to exist. However, the obstruction or overstimulation of the vessels can lead to cardiovascular problems such as diabetes, heart attacks, strokes and hypertension. Rather than having hypertension, it is imperative that we put our bodies at ease to quell the tension that could lead to death.

Hypertension is basically high blood pressure and too much force exerted in the arteries can damage the blood vessels and internal organs. The blood pressure game has been problematic because it is dependent upon numerous experimental pharmacological manipulations. For example, there are at least ten classes of blood pressure medications and over 100 types of drugs. It is a game and research has shown that ethnic groups may respond differently to each type of medication.

People of a darker hue may have a genetic propensity to have hypertension, and this may explain why some medicines work and some are ineffective. Melanin is a complex biopolymer, and it absorbs energy. In addition, diet and family history may have led to a proclivity for the biological system to be overexerted and revved up too much. Controlling the diet and mental state are ways to ease the tension without depending on drugs. It is not easy because the person must be trained to not allow the stress of life to negatively impact overall health. Consuming less salty foods, meditation and breathing exercises are ways to begin the process of controlling hypertension.

The body is the conduit to keep us in tune to a higher reality, and there are a multitude of neurochemicals or neurohormones that facilitate this mental process. Pigmentation or the presence of melanin in many regions of the body and the association of melanin with the biosynthesis of stress hormones are strong points to consider. The odd places it is located can impact the absorption of salt and other elements that can increase tension in the body's regulatory systems. Diabetes, for example, is a controllable disease, and it fully manifests when people are not in control of the sources of food that go into their body. An imbalance in glucose levels can cause high blood pressure, and high blood pressure destroys the blood vessels that feed the organs. Melanin is found in the blood vessels, the eye, the heart and kidney, so melanin-dominant people will have a higher morbidity and mortality rate from diabetic complications.

On a chemical level, there are neurohormones in our body that can collectively work to keep our minds at peace. Meditation is a way to find peace and to reach a state of consciousness that can lead to blissful life experiences. If we know melanin can absorb energy, we need to be conscious of the food, the thoughts, and the people we associate with because it has the potential to impact overall health.

Drug Interactions (Natural Highs)

A common link and biochemical explanation of what can naturally make you feel high is the neurotransmitter dopamine. Dopamine is known as the euphoria molecule, and when it is elevated, it enhances mood. Therefore, the drugs or external agents that can boost mood basically affect the transmission of dopamine in the physiological system. This conception should not be complicated to understand. However, it may be difficult to comprehend how drugs such as stimulants and depressants can both alter dopamine levels and influence a person's well being.

Stimulants drugs would include caffeine, cocaine and amphetamine. Each drug could have a different effect on the dopamine system, but the overall influence would be to boost mood. Ethanol or the alcohol you drink for pleasure is classified as a depressant. Although alcohol is a depressant, it has an ascending limb to the blood alcohol curve that can provide a "buzz" and make a person feel euphoric. In addition, alcohol has ubiquitous effects and dopamine would not be the only system it alters. Alcohol is commonly used in many societies since it is legal beyond a certain age. It is much easier to access than the psychoactive drugs and commonly abused drugs such as cocaine, methamphetamine, ecstasy and marijuana.

There are some people who cannot tolerate these external agents or consume alcohol for a host of reasons. Alcohol, for example, is not a necessary nutrient or an element in life needed by the human body. The body has all the natural elements to elevate mood and this is tantamount to getting naturally high on your own self. Yes, natural highs are real (Cass and Holford, 2002) and they exist without alcohol or other drugs.

By maintaining a focus on dopamine, it is important to understand that there are two main pathways in the brain that produce melanin. There is the nigral-striatal and the mesolimbic pathway. The former is for movement and the latter is involved in emotional behavior. The point to emphasize is that movement and mood have much to do with a person's outlook on life. Dancing, rhythm, music and any psychomotor task can affect a person's mental state in a positive direction, and this involves the nigral-striatal dopamine system.

People who do not exercise have a poorer outlook on life and they may be more prone to explore external agents such as drugs. The drugs to make them feel good stimulate the dopamine to influence the mesolimbic pathway which leads to an area of the forebrain that is highly associated with addiction, the nucleus accumbens. Therefore, the reinforcing properties of many drugs of addiction or social behaviors involved in addiction can stimulate these brain areas. The result is dramatic, and it explains why some people can develop addictions to television, sex, gambling and shopping. All of these events can stimulate the brain and make a person feel "high" or just happy.

The brain areas we just mentioned that make dopamine are associated with neuromelanin. The neuromelanin in these specific brain areas are what give pigment to these specialized brain cells. If the cells were not pigmented, the chemical dopamine would not be produced and this would create havoc on the physiological system (e.g., Parkinson's Disease). Likewise, too much dopamine could also wreak havoc on the human experience (e.g., schizophrenic-like symptoms)

There is power in these pigmented neurons, and the evidence is in the display of human behavior. If melanin can be found in the nasal passages, the oral cavity, the inner ear, the internal organs, the eye, the hair and most visibly the skin, we can understand the power of this molecule to enhance a person's extrasensory perception. The entire body is needed to reach a higher state of consciousness, so the more melanin in all of those areas can heighten what it means to be human.

Neurosensitivity (Enhanced Sensory-Motor Network)

The vitamin D concept was reviewed before as not a vitamin but as a steroid hormone (Moore, 2002). The term soltriol was created by Stumpf (1988) to better describe what we can call this entity that is stimulated by the sun and derived in the skin. Without soltriol, the body can experience a host of maladies that can lead to overall poor health. The feeling we have from a bright, sunny day is the impact of soltriol stimulating the sensory-motor network in the body.

Furthermore, we cannot take lightly the feelings we get from dancing, feeling the beat and moving to a rhythm. There are some people who cannot experience the rhythm or feel the beat. All melanin-dominant societies throughout the planet, however, have the drum as the organizing component of their civilized experience. The drum is a source of communication and the body must be prepared to receive, digest and produce the outcome from the drumbeat. Music with the drum as a base can enhance the sensory-motor network and provide structure to a civilization. When several groups of people sing and chant rhythmically together as one, the entire experience can change the way the participants as well as the outside agents feel. With no drum beat present, we can see the development of barbarism in societies controlled by melanin-recessive people.

In the brain, the skin and the internal organs are melinated structures that absorb and process light to impact our state of health. The semiconductive properties of melanin are what help the human experience to be enhanced. It creates a coordinated sequence of events to keep the skin looking fresh, the brain interactive, and the internal organs connected like the planets in the solar system. Each planet has a role to keep the solar system in order and each internal organ in the body keeps the entire body connected. The sun is the generator for the solar system and the entire human body. The presence of melanin is the conduit between the material and spiritual forces in the universe (Moore 2004), and it is clearly documented how melanin-dominant societies have always viewed a harmonious relationship with nature. To conclude, meditating to be one with nature, eating sun-enriched foods and exercising are the keys to enhancing your sensory-motor network.

CHAPTER 3

MELANIN AND NERVOUS SYSTEM DEVELOPMENT

Melanin is a self-organizing master molecule.

As a biological entity, humans were created out of the womb of darkness from a tiny cell. From the microcell to the macrobody, it is amazing how the embryological development unfolds. For example, the differentiation of organs and physiological systems are dependent upon chemicals that can ensure the proper migration of cells to form the entire body. Most cells in the body can regenerate, but nerves basically lack this capability. During embryological development, however, there is a massive growth of brain cells called neurogenesis. The rate of neurogenesis and/or survival of newborn neurons could be altered by factors like hormones (Gould et al., 1992) which genetically influence a melanin-dominant body. In the human body, we can look more deeply into the impact and power of pigmented cells in the formation of the nervous system.

For a simplistic review, the nervous system is made up of the central nervous system (CNS) and the peripheral nervous system (PNS). The CNS is primarily the brain and spinal cord, and the PNS consists of the autonomic nervous system and the somatic nervous system. The adult human brain is thought to have all of the brain cells and no additional growth as we age; however, recent evidence has shown that neurogenesis can occur in adulthood. There have been numerous reports of adult neurogenesis in the neocortex and other regions, including the amygdala, striatum, hippocampus and olfactory bulb (Gould, 2007). However, the number added is small in proportion to the brain.

The connection between melanin and brain development was previously discussed (Moore, 1995; 2004), so to add to the discourse, we will delve further into the provocative nature of how melanin-derived cells are associated with disease states such as leprosy. Since the bible has numerous references to people turning from black to white as a curse (Leviticus 13:45-46) as well as references dealing with Moses and Noah, there was something historical that relates to this discussion. Even if you do not follow the bible as a source of truth, it does not matter. What does matter is the science behind what happens in leprosy that connects the nervous system and the skin. We will return to this diseased state later, but first we must understand the biology of blackness.

For a simple review, although external and internal melanin are found in different body sites, both are embryonically derived from similar cells. After fertilization, the zygote will transform through several stages of cell replication. The cells rapidly divide during the early stages of development to form three distinct cell layers. These layers are the endoderm, mesoderm and the ectoderm, and they are collectively known as the primary germ layers (we are specifically interested in the ectoderm). The ectoderm is composed of three regions: the prospective neural tube; the prospective neural crest; and prospective epidermis. Figure 3 shows the derivatives of the neural crest cells.

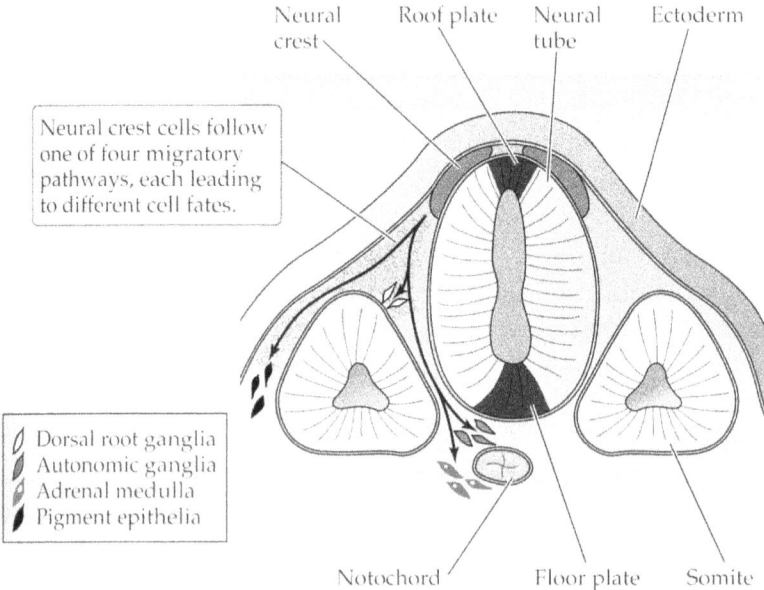

Fig. 3 - Cross-section of the cranial neural tube and the neural crest.

Melanocytes are cells that produce melanin (see Figure 4), and they are derived from the neural crest.

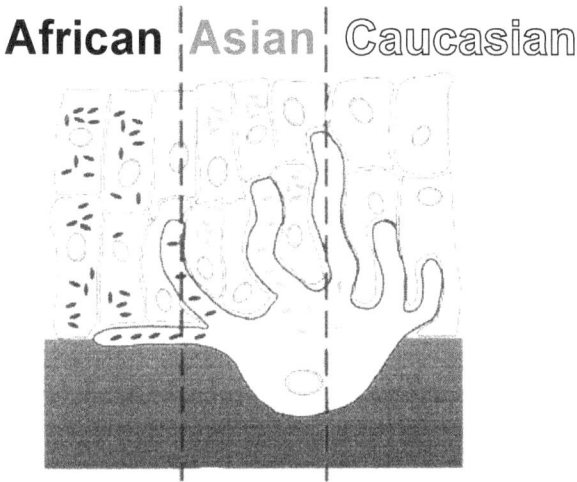

**Fig. 4 - Melanocyte and ethnic variations
in the distribution of melanin.**

Since melanin is associated with the distribution of numerous types of cells to other destination sites in the body, it is critical that we comprehend the migration of specific cells to their destination. The neural crest pinches off from the sheet of cells forming the neural tube, and these neural crest cells are lined up alongside the dorsal side of the developing embryo. According to Breedlove (2017), it is precisely because these cells lie outside the busy activity in the neural tube itself that scientists can keep track of their migratory paths. A dramatic activity of the neural crest cells is their extensive migration to take up residence in disparate parts of the body to take on a remarkably diverse range of functions, including those of the PNS. Neural crest cells migrate along well-marked pathways to reach their destinations, and the fate they take are often due to gene regulatory influences they encounter on the route and to the journey's end (Simoes-Costa and Bronner, 2015). When we speak about genes and variation from one individual to another, we can predict the migration routes and fate of neural crest cells. In *Pigment Power*, it is emphasized that *melanin-dominant people have different gene regulatory patterns for these pigment-oriented embryonic cells when compared to melanin-recessive people.*

Neuronal Fate

Originating in the PNS, the neural crest cells are not specified before they begin their migration to reach a particular goal and take on a particular fate. Animal studies have shown that if the cells are experimentally moved from one part of the rostral-caudal axis to another, they will take on the migratory route and the neural fate that is appropriate for the new position (Le Douarin, 1980). They may be guided along a migratory route by local cues along the pathway for their final destination.

For example, the neural crest cells that contribute to the ANS take on two rather different fates that we can readily detect. Some will become the postganglionic neurons of the sympathetic nervous system, forming those sympathetic ganglia that will receive innervation from a subset of spinal motor neurons, and become noradrenergic (to release norepinephrine (NE) as a neurotransmitter). Other neural crest cells will become postganglionic neurons of the parasympathetic nervous system and therefore be cholinergic (to release acetylcholine (ACH) as a neurotransmitter).

Maybe all neural crest from head to tail in the developing embryo has populations of both types of cells: those fated for NE and those for ACH. If so, then maybe in the normal case one population dies off in some regions of the rostral-caudal axis, and the other dies off in other regions. In that case, then the transplants simply switched which population would die. This was a difficult possibility to rule out from transplants alone, but a series of *in vitro* studies made it clear that neural crest cells do indeed follow local cues to follow particular pathways and their fate is determined by the combination of factors they encounter en route and at their final destination (Takahashi, Sipp and Enomoto, 2013).

Defects in Destination

As far as we can tell, of the nearly 100 billion neurons in the human brain, no two are exactly alike in appearance, so they are probably not alike in function, either (Breedlove, 2017). In the vertebrate neural crest, we know a population of cells can migrate long distances to assume a wide variety of fates. We also know that neural crest cells can sometimes "dedifferentiate," returning to an earlier, more plastic stage such that they can follow another fate. This fate can have an impact on diseased states.

On the level of importance in the CNS, a defect in this early development can cause severe neural tube defects such as spina bifida (spinal cord defect) and anencephaly (brain defect). The latter defect is associated with the major health crises in Brazil concerning the Zika virus. To combat neural tube defects, scientists recommended extra doses of folic acid before and during early pregnancy.

The Zika virus that exploded in 2016 supposedly came from mosquitoes. Since we know ethnic weapons exist and Brazil has a population of highly melanized people, it is not too far to mentally stretch how this disease just happened to explode in the manner that it did. There are historical reference points with viruses such as AIDS, so we know the manipulation of cellular components in the lab can wreak havoc when exposed to the human population. Diabolical scientists who are funded to explore population control techniques (Marrs, 2015) can exploit melanin-dominant biological systems. Although melanin can be used to enhance cellular activities, it can also be manipulated in the laboratory to cause death and destruction to the body (Harris and Paxman, 1982). Let us look at a diseased state called leprosy to get further clarity.

Leprosy (Hansen's Disease)

For thousands of years, a disfiguring disease inspired fear and repulsion. Long known as "leprosy," it is characterized by pale lumps on the skin caused by peripheral nerve damage that eventually leads to numbness, especially of the face, hands and feet (Breedlove, 2017). Being called a "leper" became the same as being identified as someone to be avoided and reviled. Because of the stigma attached to that term, today the disease is called Hansen's disease, after the Norwegian physician G. Armauer Hansen, who in 1873 identified the first bacterium to be identified as the cause of a human disease (Irgens, 1984). The invading agent was called *Mycobacterium leprae*.

To comprehend this connection, understand that there are neurons and glial cells in the nervous system. There are more glial cells than neurons (a ratio of 10:1) and some of the glial cells differentiate into cells that make the covering or insulation around axons. The covering is called myelin and this cellular substance is important for helping to properly conduct nerve impulses. In the CNS, oligodendrocytes have a fate to produce myelin. In the PNS, Schwann cells have a fate to produce myelin. It is the latent immune attack on Schwann cells that creates the problems linked to Hansen's disease in the skin and the nervous system.

For example, some of these neural crest cells differentiate into Schwann cells, and this decision to take on a Schwann cell is induced by contact with axons (Takahashi et al., 2013). The axons express a signaling molecule called neuregulin, which binds to ErbB receptors on Schwann cells to induce them to myelinate. Note that ErbB are a family of membrane-bound receptors that respond to neuregulin.

Sometimes when injury causes a withdrawal of distal nerves, the abandoned Schwann cells may dedifferentiate (Adameyko et al., 2009), and may then resume migrating. Such dedifferentiation of Schwann cells is a critical link in Hansen's disease. *Mycobacterium leprae* initially infects Schwann cells, where it replicates for years until the immune system recognizes the invader and initiates acute inflammation. The swelling in the nerve within its sheath damages the axon (Britto, 1998), which retracts, so Schwann cells lose contact with the axon. Freed from the inductive signal that kept them Schwann cells, the cells dedifferentiate to their former migrating stage in the neural crest as melanocytes going to the periphery of the body. The infected cells then migrate to muscles and skin, where they are induced to become melanocytes (Masaki, et al., 2013). The infected melanocytes trigger reactions in the skin that produce granulomas, the inflammatory swellings that mark the disease. As shown, this bacterium can cause alterations in cellular development, create deformities, and cause destruction to the human body, inside and out.

Nobles and the Biosemiotic Hypothesis

Related to the last section on Hansen's Disease, we must ask different questions about cellular systems that control both skin sensitivity and brain development. We say different questions because seldom do we have scholars who can think out of the box to create a paradigm shift on complex topics related to melanin. Other well recognized black psychologists (Bynum, 1999, 2012; King, 1991; Kambon, 1992, 1998) have incorporated melanin into their literary work by describing the importance of melanin on brain development. Along with his fellow African-centered scientists, Dr. Wade Nobles has been the master theoretician.

On an international level, Dr. Nobles is a revolutionary psychologist who has been at the forefront of the discussion on melanin as an integral component in the life of the black human. As a profound black psychologist and theoretician, he has expanded the limits of our consciousness by returning to our spiritual source as a hub to create a new humanity. A collection of his writings (2006) was shared in his book, *Seeking the Sakhu*, so that younger psychologists and scholars would have a source document of the critical ideas associated with his work as a significant contribution to the emerging field of Black Psychology. The previously mentioned research on the cause of Hansen's disease does not directly relate to his research, but the debilitation of the disease can shed light on how extrasensory perceptions can be enhanced.

In a collection of his work as far back as 1976, Nobles (2006) reflected that, "As long as the Black researcher asks the same questions and theorizes the same theories as his White counterpart, the Black researcher will continue to be part and parcel of a system which perpetuates the misunderstanding of Black reality and consequently, contributes to our degradation." P. 70. The conclusion of this chapter will address this profound statement as well as Nobles' thoughts from an African-centered neuroscience perspective.

On the topic of melanin and nervous system development, he has shared some provocative thoughts about the predominance of melanin as more than an added ingredient, which refines the CNS and produces a highly sensitive sensory motor network. A rethinking of the relationship between melanin and the CNS suggests that melanin and its structural development parallels the development of the nerve cell into the CNS. Nobles suggests that the relationship between the melanocyte and the nerve cell, both developing from the neural crest, is the basis for a complimentary sensory system network, which has a dichotomous feature. As we described in Hansen's disease, through cellular specialization, one component of the sensory network evolves from and through the melanocyte cell and the other component evolves from the nerve cell.

Nobles mentioned biosemiotics law as a connection here to respond to cellular conflict, such as the experience in Hansen's disease. Biosemiotics encompasses all living systems from the cell, over bacteria, fungi, plants and animals to humans as sign producers and interpreters. Concisely, signs are the basic units for the study of life. For Nobles, self-realization or inner self consciousness and the recognition of alien essences (conscious of other) are conflicting intentions. To overcome the conflict when the signs do not match, the next evolutionary step in the phylogenesis of life was achieved. This step is represented by what the geneticist calls "the diploidy of chromosomes." It is suggested that via this diploid cellular association one of the primal units could attend to self-realization and the other could attend to fusion with alien essences. This new inner polarization is necessary in order for the subjectivity of the organism to communicate with other alien entities while simultaneously realizing its own self.

Since cells from the neural crest eventually influence behavior, one can extrapolate behavioral enhancements from this biosemiotic process. When the signs are not balanced and interpreted properly, the system is challenged. Harmony is the orientation in nature and harmony is the glue for life. Fundamentally, the numbers and numerical proportions underlying harmony were the "chemical bond" uniting everything: music, math, art, architecture, astronomy (Finch, 2001) and for this topic, biology.

From biology to psychology, Nobles made the following assumption nearly four decades ago that, "If it is true that the phylogenetic process of self-realization is related to the establishment of consciousness, then we could suggest and understand that the consciousness of the melanocytic people and the non-melanocytic people would subsequently differ. If this is proven to be true, it would also help in the understanding that those of the species with less developed essential melanic systems would emphasize in their consciousness an orientation or dependency for outward validation. Likewise, one could expect that those of the species who have a truly balanced sensory system would emphasize in their consciousness a synthesis or union or complementarity between one's essence and external things. Parenthetically, we note with interest that in the areas of the world where the largest number of the high-melanic-possessing people are found, so called paranormal powers, psychokinesis and precognition, generally referred to as voodoo or witchcraft, are also found." pp. 65-67

Conclusion

Melanin is an organizing molecule that affects multiple cellular systems. It keeps the body healthy and it can help ward off illnesses and fight disease. Therefore, one could say melanin is important to boost the immune system to keep the human body healthy.

The use of the term "immune" system is a misnomer because people are not immune from anything. The 2020 pandemic from a specialized coronavirus shows how vulnerable the human body is to pathogens. To be specific, a person may have a defense mechanism system set up to fight against insults to the body. To use the term immune system, however, there is an assumption that it is a thinking system. It can protect immediately, it can attack the body's own cells, or it can respond to foreign agents circulating in the body years away from the initial attack on the body. As it thinks to protect us, it can wreak havoc on the total body response. Hansen's disease is the perfect example.

The fact that you have a disease that connects the brain, melanin and the formation of the nervous system sheds scientific evidence on the power of pigmentation in the human experience. If it is malfunctioning and deficient, we can see the manifestation of diseased states. In contrast, when pigment power functions at full capacity, it greatly enhances the human experience for extrasensory perception. For extensive studies, search the course work offered by Jewel Pookrum at the J.E.W.E.L. University of Immortal Sciences for Immortal Living (www.juis.education).

CHAPTER 4

SUNCONSCIOUS

A sun-dripping consciousness can heighten your reality.

One Spring while visiting Europe, I had one of the most fascinating experiences that you probably would not believe. I ate a meal in the cave while trekking along the hillsides of Europe. To be specific, I ate an extravagant lunch at a restaurant which was located inside a cave in Tours, France. This unique experience has caused me to seriously reflect on the impact the environment has on our consciousness. Historically, living inside a cave on a daily basis was protection against the harsh elements of Europe. After spending time in the cave, I realized the significance of why a lack of sunlight was detrimental to physical development as well as mental consciousness. From the dark cave in cold Europe to the heat of equatorial Africa, there is a change in consciousness. As we reflect on this change, we call it the "sunconscious."

When we reach the month of July, we have already passed the summer solstice around June 21. July and August usually provide the most intense heat in the western hemisphere and near the equator, so let us reflect on the fact that the summer months are when the heat is on. In the beauty of your blackness, however, is the natural protection against the harmful effects of solar radiation. As African people, we should be living in glory as the pigment known as melanin daily energizes our mind, body and consciousness. The summer presents a different "sunconsciousness" and it is all stimulated by "soular" power.

Our ancestral roots are in Africa, and the intense heat of Africa has caused our genes to manufacture the smooth coat of beauty that resides on the outside of our body. Melanin is the biological entity that allows us to live in peace and harmony with the sun. The sun can be dangerous to our genetic structure, but life would not exist without the sun. Therefore, the existence of the sun, the presence of melanin, and the uniqueness of the African spirit can combine to elevate the consciousness of a people.

What evidence do we have for this sunconscious experience? First, the beginning of all civilization began in Africa and the right combination and elements obviously led to the origins of civilization from the continent of Africa. In contrast, the mind of a people living in a colder climate will not be focused on building monuments when they are living in ice age conditions. Secondly, think how positive you feel when the sun is shining versus the feeling you have on an overcast and gloomy day. Yes, the existence of the sun, the presence of melanin in the human body and the uniqueness of the African spirit have shown us the manifestation of God in human form.

I have written on melanin from several perspectives, and this treatise will shed sunlight on why I have come to the conclusion that emotions are strongly connected to the presence of melanin. Melanin absorbs, retains, and emits energy and this physical capacity provides the basis for linking a physical element (melanin) with a nonphysical entity (spirit or consciousness). The problem with many people is they are unable to channel their emotion properly to keep stable connections between people. Most arguments and disagreements between people are usually a result of no "emotional" connection or a lack of emotional intelligence. In other words, the vibes are off and not in sync. Many human problems are due to simple and trivial matters that mean nothing in the grand scheme of things. Reflect upon how people of African descent are often caught up in emotional issues that disintegrate the family, love and trust between one another. Think about how much does not get done in the African Diaspora because we get caught up in feelings rather than the issue that needs to be addressed (Ridley, 1980).

Too often, melanin-dominant people are relegated to the bottom rung of social stratification on a world-wide scale. We must make changes to this shameful outlook. I did not say analyze our problem because we have overanalyzed our dilemma with no solutions that have addressed the dismal state of African progress for centuries. I believe our empathy and feelings have been our downfall as we have come in contact with people who have little compassion for humanity. It is a double-edged problem, however, because too much emotion can cause you to be taken advantage of and too little or the wrong use of emotion can lead to destructive relationships. It may be recondite to consider how pigmentation could be responsible for the emotional issues I am raising but be patient with this presentation. Let me return to the cave to endarken you, and then we will come out of the cave and see the importance of the sunlight that nourishes all life.

Melinated people are sun-driven and the heat of the sun can energize the human spirit in a positive way. A lack of sun, however, can lead to physical abnormalities and a lowered mood state. Some people literally cannot deal with the heat, and this heat can sometimes be in the form of stress. Therefore, the combination of stress and the heat of the sun can dramatically influence the moods of people and the presence or absence of melanin can affect the way that energy is absorbed. If we look at melanin as an absorber of energy, then we can see how melanin-dominant people can adapt to and tolerate soular energy better than melanin-recessive people. If thoughts are energy, then we can see how the presence of melanin can function as an absorber of conscious thoughts to perhaps make darker-skinned people more spiritual. It may be more of a challenge to convince you that melanin and consciousness are connected; however, I believe a focus on the physical nature of melanin and its role as a pigment will endarken your understanding (Figure 5).

SUNCONSCIOUS

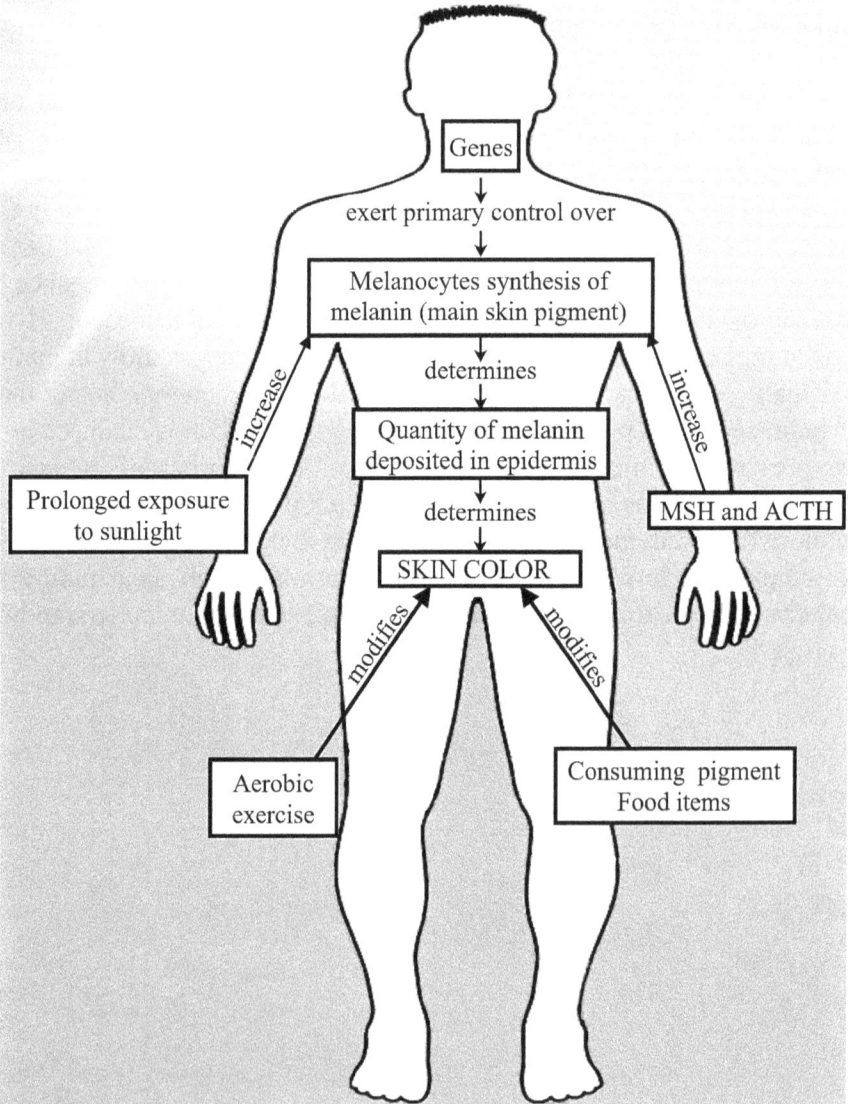

Fig. 5 - Genes determine an individual's basic skin color by controlling the amount of melanin synthesized and deposited in the epidermis. Beyond the sun, other factors can increase and modify pigmentation.

Sun Depletion

When I was in the cave for a short period of time, it was cold, dark and dank. Soular energy could not penetrate the hard rock and the deeper we went into the cave, the farther away we went from light. Modern caves may have electricity for light, but ancient life in caves used other forms of light that did not last as long as a light bulb. Living in the cave meant that the nourishment from the sun was absent. No sun means no vitamin D production in the skin. As I have reiterated elsewhere (Moore, 2004), vitamin D is actually a steroid hormone and not a vitamin. The fact that it is a steroid hormone should add greater significance to how moods can change when the sun is present.

When there is a lack of sunlight, a physical abnormality called rickets can develop. Rickets is a bone deformity that can make the bones weak and brittle. Dark-skinned people who migrated from Africa were stuck in these caves during the ice age. These dark-skinned people living in Europe changed to lighter-skinned people to adapt to the sun-depleted environment. The impact was great because they changed internally and externally.

Sun Enrichment

On the other hand, a sun-enriched environment can do wonders to the soul, spirit and consciousness. Yes, too much sun can be harmful to the skin, but it is a matter of being wise and not standing and baking in the sun too long. Moreover, using sunscreen lotions are supposed to be effective and it is important to avoid the sun when it radiates its most intense heat. We cannot live without the sun, so that it is not even a discussion. What the sun does is help establish our biological rhythms. The presence of this massive soular generator keeps all life in rhythm, and we all feel much better during the summer months when the sun is shining more and longer. As soon as the seasons change and we move toward the fall and lead into winter, some people are really affected, and they experience winter depression or Seasonal Affective Disorder (S.A.D.).

Working in closed environments without natural light can alter moods. Working in an environment with artificial and fluorescent light can impact moods. We need the right amount and the right exposure to sunlight to keep our biological systems balanced. Therefore, a sun-enriched environment is critical for maintaining good physical and mental health.

Soular Power

The soul had always been an integral part of the cosmology of ancient Africans (Akbar, 1994). From Ethiopia to Nubia to Kemet to the migration of Africans all over the continent, the soul was incorporated into African philosophy. For example, the first two of the seven divisions of the soul in ancient Kemet were Ka (the body) and Ba (the breath of life). The Kabala in Jewish mysticism and the Kaba stone in Mecca are nothing but extensions from the soul concepts presented in Africa. In addition, the sun or Ra was praised as the first monotheistic approach to God. When we say soul brother and soul sister today, it means that we are consciously reflecting on our history and what it means to be connected to all of humanity.

Melanin is the black gold of the sun. It is the element in your body that keeps you lively and rich. Soular power is generated from the energy transforming capabilities of melanin in and outside of the human body. As I sat in the cave, I could see and feel the soul being zapped out of me. The absence of the sun seemed to diminish all vitality. I became cognizant of the importance of melanin as the humanizing factor in that environment. From this experience, I believe the loss of melanin in those individuals living in the caves and hillsides of Europe set the stage for aggressive and control-oriented thinking. A lack of empathy for others turns humans into complex animals, and I felt the loss of the essence of blackness in the dark cave, and it gave me an unnatural way of viewing the world (see Bradley, 1978).

I came out of the cave in search of soular power because without the sun, life seemingly ceased to exist. I am convinced that without soular power, prehistoric humans lost their minds in those caves. Interestingly, modern scientists are debating where the mind is actually located. We are not getting far by assuming consciousness only emerges in some way or other from brain activity. As an African-centered scholar, I must turn to the ancient African philosophers for answers. They believed the mind or consciousness existed in the cosmos as fundamental space, time and matter. From the Kybalion and the Seven Principles of Life (Chandler, 1999), The First Principle called The ALL or The Mind was everywhere.

Modern science teaches us that melanin is everywhere in and outside of our human body, in inanimate objects such as fruit and in the cosmos in the form of dark matter. Since scientists still cannot defini-tively state where consciousness resides, we might need to change our thinking to the reality that consciousness is fundamental to life. The historical evidence suggests that living in a cave attributed to the loss of pigmentation, and subsequently, a disturbed mind. What existed after this cave experience was individualism and an unnatural way to per-ceive life. On the other hand, people who meditate and reach divine universal consciousness and penetrate the core of their mind find that the light of consciousness shining in them is the same light that shines in you and every other human. The physical properties of melanin and the energy transforming capabilities of melanin are the source of this divine universal consciousness.

In conclusion, understanding melanin and its role in the mind may be the next frontier in science. When the ancient mystics spoke of the divine, they were not speaking of some supernatural being who rules the workings of the universe; they were talking of the world within. Scientists have yet to explore the realm of the "deep mind" within, and the experience in the cave made me aware of how far removed we are from our individual souls, collective mind and cosmic consciousness. When humans lost their melanin, they lost their minds.

CHAPTER 5

ELECTROMAGNETIC ENERGY
AND MENTAL HEALTH

The right light will light your light.

In the scientific literature, there is evidence that there are genetic differences in how melanin-dominant people process energy differently when compared to melanin-recessive people. The exploration of this evidence will be explored in this chapter. With this understanding, we can begin to comprehend the power of pigment.

Our bodies respond to a large electromagnetic spectrum of energy that exists in the entire universe. Light and sound are two major factors of this vast spectrum in which melanin must be present for the body to process this external energy. Throughout the body, the pineal gland, the skin, the eyes, and the hair function as melanized transducing components of the human body to absorb and transform this energy into a form in which the body can use. There is magnificent power in light and sound energy, and it can be harnessed to benefit our life (Ott, 1975). The electromagnetic spectrum is vast but visible light is all we see with our physical eyes. Looking into the sky, we can see there is no shortage of light on this planet, but the visible light we see is only 1% of the full electromagnetic spectrum. How light is used determines the outcome of the human experience. Similarly, music can be a weapon used against you or it could be used by you on a personal level to heal. Essentially, our eyes and ears are key points of entry for electromagnetic energy to enter the body.

Light Energy and Melanin-Dominant People

Sunlight triggers the release of soltriol (Vitamin D), serotonin, nitric oxide, endorphins and a host of other natural compounds in the body. For human beings, proper exposure to the sun can reduce the risk of prostate, breast, colorectal and pancreatic cancers. Furthermore, we know sunlight can improve circadian rhythms, reduce inflammation and dampen autoimmune responses. There are people who have allergic reactions and the sun can make the skin worse, but overall, the sun provides the necessary energy to trigger positive mental health.

Light is therapeutic for the human experience by activating mechanisms in the skin, the eyes and the endocrine system (Figure 6).

MULTIDIRECTIONAL COMMUNICATION INTERFACE (MCI)

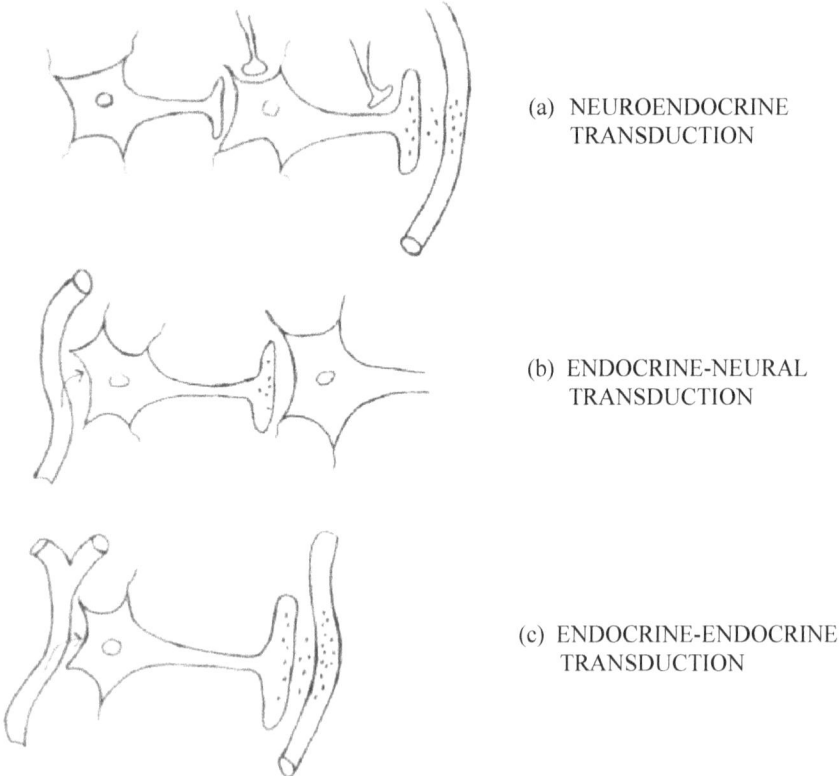

(a) NEUROENDOCRINE
    TRANSDUCTION

(b) ENDOCRINE-NEURAL
    TRANSDUCTION

(c) ENDOCRINE-ENDOCRINE
    TRANSDUCTION

Fig. 6 - Multidirectional communication between the nervous system and the endocrine system.

Since light is energy, light can stimulate biological systems and cause a host of positive health consequences. In contrast, the absence of light for humans can also trigger death. When high intensity light hits the surface of the skin, this radiation has a powerful influence to make the cells in the skin produce melanin. The melanin absorbs light energy, and it can serve as a neutralizer of toxic substances, so the intensity of the light does not cause DNA damage. This protection is vital because DNA damage can lead to skin cancer. As the largest organ in the body, we can see the importance of exposing the entire body to the sun to maximize the health benefits. You feel so much better when nourished by natural light, and the skin provides full body immersion.

Along with the skin, the eye is like a small filter for light to enter the brain. The eye is uniquely structured to channel light through a pin hole (pupil) to the back of the eye where the photoreceptors are located to process light. The photoreceptors are called rods and cones, and they are embedded in a fine layer of melanin in the retinal pigment epithelium. To understand the importance of the melanin in this location, those individuals who have an absence of this melanin have a disease called retinitis pigmentosa. The absence of melanin in the back of the eye renders the person virtually blind.

The melanin in the eye attributes to mental health in a variety of ways. First, melanin absorbs the ultraviolet radiation to prevent damage to the sensitive photoreceptors. Secondly, the absorption of the energy can trigger the regulatory transmission of natural chemicals in the eye region such as dopamine and melatonin. Dopamine is associated with euphoria, and melatonin is also important for impacting mood. It is well documented that there are both dopamine and melatonin receptors in the eye, so the presence of these chemicals in the eye can indirectly contribute to mental health. When it is stated that "the eyes are the windows to the soul," we can relate this to the powerful effect of how brain chemicals can be stimulated by light (Figure 7).

Pinealocyte as a Photoreceptor

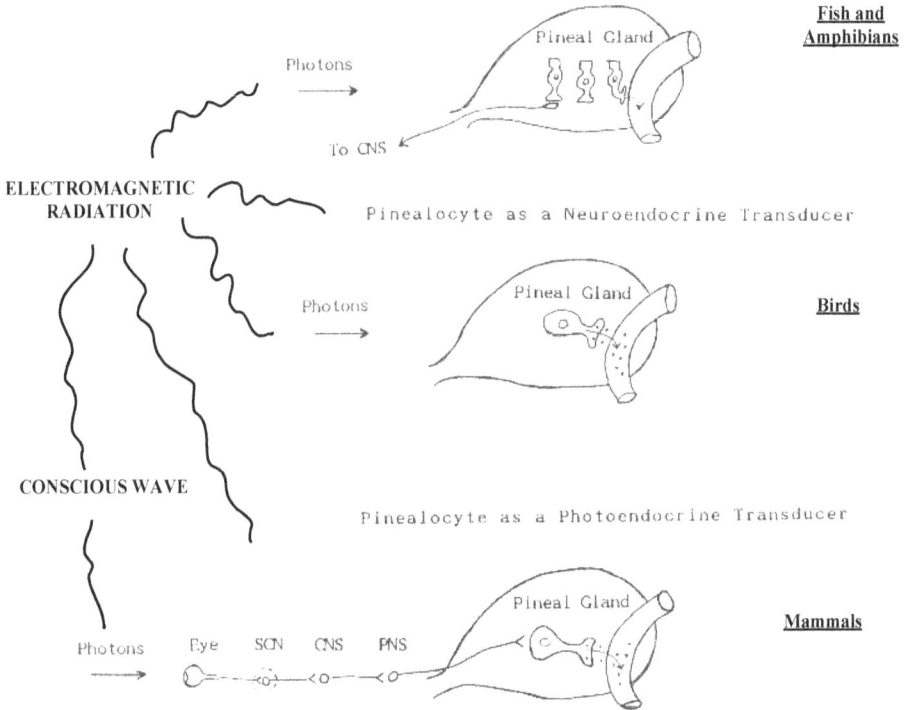

Fig. 7 - The pineal gland is an intermediary link between the
environment and the endocrine system.

The pathway from each eyeball leads to the brain, and the
pathway is intricate and revealing. Surprisingly, this multilevel pathway can reach the spinal cord (superior cervical ganglion), the pineal
gland, the hypothalamus (suprachiasmatic nucleus), and the occipital
cortex as the final location where vision is processed. As a result of
these connections, the endocrine system is the output system from this
solar generated input. Endocrine implies the secretion of chemicals
inside the body that regulate multiple aspects of behavior. What light
does is entrain all of the biological systems in a rhythmic pattern. The
circadian clock, for example, relates to the 24-hour biological rhythm
that is in tune with the rising and setting of the sun. Light is the
zeitgeber for keeping our body and mind stabilized. In other words,
light is the external cue that entrains or synchronizes an organism's
biological rhythms to the earth's 24-hour light/dark cycle.

The Science Behind Light and Health

There have been numerous melanin-recessive scientists who have revolutionized our understanding of electromagnetic energy and how it can be beneficial to the human experience. For example, Nikolas Tesla has developed the technology for us to have wireless communication and other ways to harness electrical energy. Wilhelm Reich developed an apparatus (The Orgone Machine) to harness the universal life force to alter people's mood. Both Tesla and Reich were geniuses, but they were not fully supported by the establishment when they were alive; Actually, they died poor. A third scientist to mention is John Ott. He lived a long life up until the age of 91 and he was an expert in light and photography. The topics addressed by Ott would also not be supported by the establishment.

For example, Ott (1975) considered artificial light to be pollution. Henceforth, he dedicated his life and research to making people's lives better with full spectrum light. Ott was not a medical doctor so we can see how the medical establishment could potentially dismiss his research. He did, however, have an honorary degree of Doctor of Science due to his extensive research and experimentation on plants, people and laboratory animals.

If you explore the work of scientists like Ott, Reich and Tesla, one can see how they were addressing health issues that could change the direction of the capitalistic enterprise we live in. For all three of these researchers, they received limited support from big companies because their work could change the economy. Since capitalism is based on profit over people, health is of limited concern. The aforementioned comment is not stated lightly to be glossed over because we can see the polemical discussion capitalist have over healthcare for all. For the campaign of the 45th President of the USA and his capitalistic contacts, one main goal was to wipe out the Affordable Care Act.

With this narrow-minded and polluted political view to discourage universal healthcare, it is very clear the capitalistic enterprise has no interest in healthcare for all.

To refocus our discussion on light along with this polluted thinking, a person might interpret light as pollution that can be absorbed by the body to have detrimental consequences. There are different types of light that can negatively impact mood, learning and overall health on a molecular level, and according to Ott, "mal-illumination" can destroy our health. We need light and life on earth evolved under natural outdoor sunlight. Our various types of artificial light sources, however, represent gross distortions from this natural sunlight. Window glass, automobile windshields and glasses can distort by filtering out the ultraviolet part of the spectrum (Ott, 1975). Because of light distortion, we can observe a poor learning environment in schools, the progressive development of cancer, and a host of health maladies that can alter a person's immune system and biological rhythms.

In terms of research on light and mental health, there are several directions that can be explored. Light plays a pivotal role in the regulation of affective behaviors. Huang et al. (2019) demonstrated in rodents that light influences depressive-like behaviors through a disynaptic circuit linking the retina and the lateral habenula (LHb). Specifically, M4-type melanopsin-expressing retinal ganglion cells innervate GABA neurons in the thalamic ventral lateral geniculate nucleus and intergeniculate leaflet (vLGN/IGL), which in turn inhibit CaMKIIalpha neurons in the LHb. The results revealed a dedicated retina-vLGN/IGL-LHb circuit that regulates depressive-like behaviors and provide a potential mechanistic explanation for light treatment of depression. This habenula is a neuroanatomical site where there is a direct connection between the pineal gland to the brain and the overall influence on behavior is significant (Figure 8).

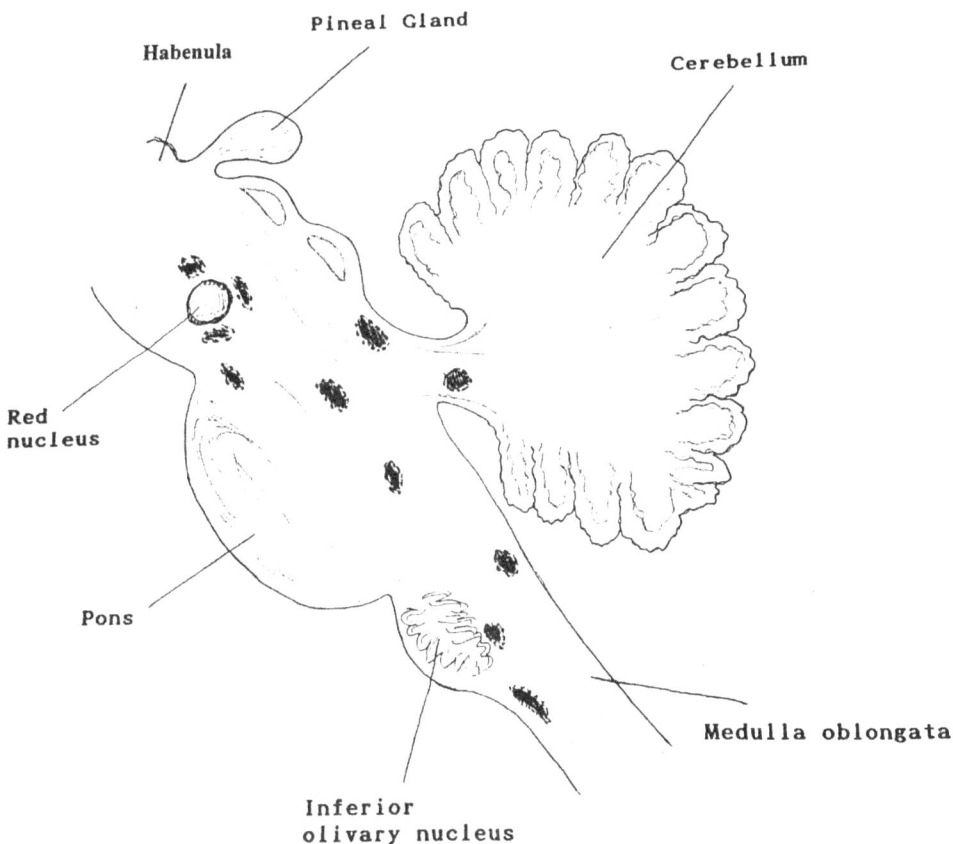

Fig. 8 - Sagittal section through the brain stem showing the habenula area at the base of the pineal gland and the neuromelanin in the Amenta nerve tract.

In Norway, Henriksen et al. (2016) studied the effects of blue-blocking glasses to treat mania. The discovery of the blue light-sensitive retinal photoreceptor responsible for signaling daytime to the brain suggested that light to the circadian system could be inhibited by using blue-blocking orange tinted glasses. Therefore, blue-blocking glasses could be a potential treatment option for bipolar disorder in a manic state.

Other researchers (Brainard et al., 1990) studied the effects of different wavelengths of light in seasonal affective disorder (S.A.D.). The aim of this study was to compare the relative therapeutic efficacies of three different light sources for treating winter depression or S.A.D. A balanced incomplete block crossover design was employed, whereby all 18 patients were randomly assigned to two out of three treatment conditions: white, red and blue light. The data suggested that a photon density of 2.3 X 10(15) photons/s/cm squared, white light has greater therapeutic benefit than red or blue light.

Disner, Beevers and Gonzalez-Lima (2016) combined low-level light therapy (LLLT) with attention bias modification (ABM) as a cognitive intervention to improve depression in 51 adult patients. LLLT with transcranial laser is a non-invasive form of neuroenhancement shown to regulate neuronal metabolism and cognition. The participants with elevated symptoms of depression received ABM before and after laser stimulation and were randomized to one of three conditions: right forehead, left forehead, or sham. Participants repeated LLLT two days later and were assessed for depression symptoms one and two weeks later. Disner et al., concluded that the beneficial effects of ABM on depression symptoms may be enhanced when paired with adjunctive interventions such as right temporal LLLT.

Pharmaceutical agents are widely used to treat mental health challenges. To combine drugs and light, Lam et al. (2016) studied the efficacy of bright light treatment, fluoxetine, and nonseasonal major depressive disorder. The patients (aged 19-60) were from outpatient psychiatric clinics, and they were diagnosed with major depressive disorder. The patients were in a randomized, double-blind, placebo- and sham-controlled, 8-week trial. Patients were randomly assigned to: (1) light monotherapy and placebo pill; (2) antidepressant monotherapy (20 mg/d of fluoxetine); (3) combination light and antidepressant; and (4) placebo pill. In conclusion, bright light treatment, both as monother-apy and in combination with fluoxetine, was efficacious and well tolerated in the treatment of adults with nonseasonal major depressive disorder. The combination treatment had the most consistent effects.

In conclusion on the topic of light and overall health, Ott (1975) researched the importance of full spectrum light. However, more recent studies have presented contrary evidence when explored over a span of time. McColl and Veitch (2001) reviewed data covering the period 1941-1999 to consider evidence for direct effects of full spectrum fluorescent light through skin absorption as well as indirect effects on hormonal and neural processes. Even though the data reported by McColl and Veitch did not reveal dramatic effects of fluorescent lamp type on behavior or health, neither does it support the evolutionary hypothesis. Without knowing the types of communities studied or the ethnicity of the subjects, the data might be difficult to interpret. However, as emphasized by Ott decades earlier, full spectrum light has been credited with beneficial effects on a wide variety of behaviors, mental health outcomes and physical health effects as compared to other fluorescent lamp types.

Sound Energy and Melanin-Dominant People

Sound is basically a wave of energy that is part of the electromagnetic spectrum. Perhaps humans could exist without hearing sound, but our entire body would still feel the vibrations of this energy. Therefore, the vibration of sound is felt by the entire body and we can break down sound into components such as amplitude and frequency. The amplitude deals with intensity or loudness of a sound wave and the frequency relates to the rate or cycling of a sound wave. On a human level, sound can be heard to create language to communicate. In this language of communication, sound can be manipulated by humans to make music. Both language and music are important for stimulating positive mental health.

The processing of sound is dependent upon the ear and the internal mechanisms of the inner ear where the hair cells are located. In the inner ear in the cochlea is where there is a heavy presence of melanin and the cells that produce melanin. Perceptions of sound are subjective, so attempting to study the effects of sound on mood are dependent upon a host of factors that are beyond the ear. During a youthful age and throughout adulthood, the ear is responsive to outside energy. Hearing loss occurs with age and studies have been conducted to investigate skin pigmentation and hearing loss.

In the science literature, it is known that black individuals have a lower risk of hearing loss than do whites. The fact that melanocytes are differentially expressed in the cochlea is the reason. This research is extending findings that darker-skinned individuals tend to have more inner ear melanin, and the cochlear melanocytes are important in generating the endocochlear potential. The manifestation of what people hear with their ear can be different; however, the data on hearing loss is difficult to interpret.

In a massive study conducted by Lin et al. (2012), they analyzed cross-sectional data from 1,258 adults (20-59 years) in a survey for pure-tone audiometric testing. Among all participants, race/ethnicity was associated with hearing thresholds (black participants with the best hearing followed by Hispanics and then white individuals), but these associations were not significant in analyses stratified by skin color. In contrast, in race-stratified analyses, darker-skinned Hispanics had better hearing than lighter-skinned Hispanics. Surprisingly, associations between skin color and hearing loss were not significant in white and black participants. These findings demonstrate that skin pigmentation is independently associated with hearing loss in Hispanics and suggest that skin pigmentation as a marker of melanocyte functioning may mediate the strong association observed between race/ethnicity and hearing loss.

In another extensive study by a different team of researchers (Lin et al., 2017), approximately 49,323 white women in the Nurses' Health Study were studied to measure skin pigmentation and risk of hearing loss. It is interesting that only white women were studied, and in this cohort, the surrogates selected for skin pigmentation were not associated with risk of hearing loss. Since it was only white women, the surrogate measures of pigmentation were hair color, skin tanning ability, and skin reaction to prolonged sun exposure.

When studying cadavers, even more evidence for the differences can be found on a genetic basis between ethnic groups. Sun et al. (2014) characterized the distribution of melanin pigmentation in the human cochlea. Human temporal bone specimens from the Johns Hopkins Temporal Bone Collection were examined. Demographic, clinical, and audiometric data were analyzed. Nineteen African American and 27

Caucasian specimens were selected. The mean ages were 64 and 70 years for African American and Caucasian specimens, respectively. At all cochlear turns, African American specimens contained significantly more pigmentation in the stria vascularis and Rosenthal's canal compared to Caucasian specimens. The authors concluded melanin pigmentation is significantly more abundant in African American than in Caucasian cochlea. This study provides a detailed description of pigmentation in the cochlea and may help to explain the observed racial differences in thresholds.

Using the same Johns Hopkins collection, Erbele at al. (2016) studied racial differences of pigmentation in the human vestibular organs. They quantified the melanin pigmentation in the vestibular system and examined racial differences of vestibular melanin pigmentation using human cadaveric temporal bone specimens. Light microscopy was used to examine specimens from 40 left temporal bones. Color images of the (1) ampulla of the horizontal canal; (2) the utricular wall; (3) the endolymphatic duct; and (4) the posterior ampullary nerve were acquired. Fifteen African American and 25 Caucasian specimens were analyzed, and the mean age was 68.8 years. African American specimens had a significantly greater amount of pigment at all four sampled locations as compared to Caucasian specimens. In conclusion, the authors state there is a greater melanin pigmentation within the vestibular system of African Americans than in Caucasians, similar to what was described in the cochlea.

Overall, these findings can help to explain the complex psychomotor skill development in combination with rhythm that is expressed in melanin-dominant cultures. The movements associated with simultaneously dancing and rapping, the spontaneous athletic moves, and the display of creative music forms come from a highly sensitized sensory-motor network of cells that begin in the brain.

Inner melanin can mute acoustic waves and make sound "feel" differently. The bass in black music is felt, and with no bass, there really is no rhythm. The syncopated rhythms of the drum in all forms of music played by melanin-dominant people has influenced music all over the world. In fact, the sound of black music pervades, invades and is fully expressed in the soul – Soul Music.

Sound needs melanocytes to be heard (Tachibana, 1999). Intermediate cells in the stria vascularis of the mammalian cochlea are melanocytes, which contain melanin pigments and are capable of synthesizing melanin. These melanocytes are required for normal development of the cochlea, as evidenced by studies of mutant mice with congenital melanocyte anomalies. Melanocytes are also needed for developed cochlea to function normally, as evidenced by studies of mutant mice with late-onset melanocyte anomaly and humans with acquired melanocyte anomaly.

Due to extensive research, how inner ear melanin is made is less of a mystery today. Melanocytes were found in the subepithelial layer of the dark cell area of the vestibular organs. These dark cells had round or spindle-shaped nuclei and clear cytoplasm with brown pigment granules which were thought to be melanin granules (Masuda et al., 1995), and this melanin formation in the inner ear is catalyzed by a new tyrosine hydroxylase kinetically and structurally different from tyrosinase (Benedito et a., 1997). In other words, melanogenesis in the cochlea, and likely in other extracutaneous locations such as the brain, is catalyzed by enzymatic systems different from, but related to tyrosinase. The combined results of the data analyzed in this chapter provides evidence for why and how people of different ethnicities can hear and feel sound differently, even in the presence of the same experience.

To conclude this chapter, we have explored the way energy can be harnessed by melanin for the body to use. The eye and the ear were the physical elements of the body that help to process electromagnetic energy. Both sound and light waves help to establish rhythm on multiple levels, so the next chapter will explore the overall impact of rhythm, from a melanin-dominant perspective.

CHAPTER 6

RHYTHM NATION

You GOT SOUL; If not you would not be in here.

Despite the world-wide influence of having #44 as a melanin possessing President of the United States of America for eight years (Coates, 2018), the social pendulum swung backwards with the presence of #45 at the helm in less than one month of his inauguration. Our nation is out of rhythm, and the continuation of oppressing melanin-dominant people via laws and legal decisions (Alexander, 2012; Blackmon, 2009; Wright, 1990) has been devastating to melanin-dominant communities in the USA and throughout the world.

A change is needed, and this chapter provides a social commentary on how power has been depleted from a movement-oriented people. Although melanin in our nervous system contributes to having rhythm, we will not directly discuss melanin. Instead, we will highlight the rhythmic swing from #44 to #45 and why Black males are a target for control. We cannot say destruction because black males are a commodity to be exploited for the corporate enterprise (e.g., think about March Madness in college basketball). Therefore, we must analyze some factors contributing to the dysfunctional behavior expressed in many African/Black males.

Many young people are misguided and lack direction in life. Without direction, there is great potential that any road chosen will guide the youth towards a path of self-destructive behavior. In the current state of affairs in this ruling patriarchal world controlled by melanin-recessive men, the most dominant threat to the melanin-recessive man is a melanin-dominant male. Too often, Black men are criminalized, vilified and victimized to eliminate competition in the Eurocentric reality of material gain (Welsing, 1991). In order to find answers and solutions to the many dilemmas faced by young

African/Black males, there is a need to find positive attributes expressed in the culture of African/Black people.

The thrust of this chapter is to focus on an ancient and positive attribute expressed in African/Black culture called rhythm. The expression of rhythm has been an influential cultural factor that has historically led to the success of melanin dominant people. In ancient times, rhythm was a very important principle for life. The reaffirmation of rhythm has been espoused for contemporary times (Chandler, 1999), and we will highlight the Principle of Rhythm as it was expressed in ancient times. However, it is the contemporary use of rhythm that is important to our discussion. African/Black males are in great need of proper paths to follow to achieve success in entertainment, athletics, education, health, fashion, relationships and marriage. This chapter is written to provide the proper ebb and flow that is needed to produce a nation and generation of positive African/Black males.

The Principle of Rhythm

There are seven Hermetic laws or principles that are known to have originated in ancient Egypt, and a detailed view of how to use these principles for contemporary life have been discussed (Chandler, 1999). One principle called rhythm is evident in the creation and destruction of worlds, as well as in the rise and fall of nations. According to the Kybalion:

> *"Everything flows out and in; everything has its tides;*
> *all things rise and fall; the pendulum swing manifests*
> *in everything; the measure of the swing to the right,*
> *is the measure of the swing to the left; rhythm compensates"*

If we train the right cadre of young African/Black males, a rhythm nation is due to prevail, but we must first make the nation-builders aware of this fascinating principle. The remaining six principles are mentalism, correspondence, vibration, polarity, gender and causation.

Entertainment

Several years ago, Janet Jackson used Rhythm Nation as a theme for one of her major musical productions. No one can deny her great skills as an entertainer, and it was part of her ancestral heritage that she was able to use Rhythm Nation as a theme. The subject matter in this book has everything and nothing to do with the talented Janet Jackson nor the content of her album/CD. To say everything and nothing in the same sentence appears to be an oxymoron, but this statement is the first rhythmic pulsation to be expressed here. Everything means that rhythm is one source of strength she was using for entertainment purposes to provide excitement to her listening audience. Nothing means that Janet, as a female entertainer, has nothing to do with this intellectual discussion on the use of rhythm as a source of strength for African/Black males. Actually, the fantastic musical rhythms expressed by Janet and the pulsating, gyrating movements of her body are the distraction for many African/Black males. As discussed later, the Black woman has everything to do with our discussion, for she is who propels the mind of men towards positive rhythmic vibrations. On the other hand, Janet as an individual has nothing to do with building a nation of strong, disciplined African/Black males. She certainly thrusts with her busts, but the thrust of this chapter is to focus on the ebb and flow of the visual distractions (the ebb) that sidetrack African/Black males and the musical beats (the flow) that help to stimulate and create.

Historically, the drum had been used as a traditional and ceremonial aspect of African/Black culture. Prior to our contact with Europeans, the drum was used to communicate in ways non-African/Black people could understand. The drum was actually banned from use during chattel slavery in the Americas because White slave masters considered the drum to be a dangerous source of upheaval. The potential destruction of our rhythmic nation began with the forced enslavement of African people. Fortunately, there is no possible way to destroy rhythm because it is tantamount to stopping a heartbeat. You may have bad health, but the heart will continue to beat. Throughout the African diaspora, we can see remnants of African culture and how rhythm has influenced music from Cuba to Belize to Brazil to the Polynesian Islands to the Caribbean Islands and anywhere else there is a large contingent of African people.

Along with the drum beat and rhythmic music is dance and we all know how important dance and movement is to African culture. Without mentioning any names, the greatest entertainers in pop culture are those individuals of African descent who gyrate, pulsate and move to a rhythm that no other group of people appear to manifest. Many cultures have incorporated dance into their cultural experience, but you know if a musical genre has been influenced by the stamp of African rhythms. As background information, the emphasis here is that African/Black males continue to set the trends in many sources of musical forms. From African American classical music (i.e., Jazz), to bebop, to rhythm and blues, to rap, to hip-hop to neo-soul, the creativity stemming from our African heritage is powerfully influential.

At some point, we must use this rhythm as a source of building a respectable nation. For example, Public Enemy's song *Fight the Power* and KRS-One 's album *Edutainment* reveal how revolution can be put to the rhythm of a beat. They have tried to stay positive as revolutionary-type entertainers, and black males should continue to stay in this vanguard.

When we talk about power and influence, not many can pass Janet Jackson's brother, Michael Jackson, the "King of Pop." Although he went through disfiguring transformations with his physical appearance, he produced an album in 1995 called *HIStory: Past, Present and Future, Book I.* The album theme indicated that Michael became more conscious of controlling his empire during the latter part of his existence on this planet. We spoke about James Brown in Chapter one, but even in death, Brown's music is still being sampled. Along with the untimely deaths of Jackson and Brown, we must mention the life and influence of Tupac Shakur and Biggie Smalls. Even in death, Jackson, Brown, Shakur and Smalls still have a global influence on music. Those that are alive, however, have truly taken pigment power to another level.

For example, the roll call of contemporary influential melanin-dominant performers with a revolutionary mission is led by Jay-Z. He was listed to have a net worth of one billion dollars in 2019. In the unique tradition of rap royalty, Beyoncé (500 million net worth) and Jay-Z have combined their talents and income. Along with their collective talents, there is Roc Nation (entertainment),

Roc-A-Fella Records (music) and Rocawear (Fashion line). One billion dollars is power, but the other influential melanin-dominant artists who have a little less swag with their 2019 net worth would be entertainers like J. Cole (60 million), Kenrick Lamar (75 million), Lil Wayne (120 million), Kanye West (250 million) and Sean Combs (740 million).

Everyone has their preference of music, but we should be choosing music that is essential for spiritual and mental upliftment. In the opinion of this author, any music that degrades humanity does not have any inspirational value. Rhythms with consciousness must be used to feed the people because the people are starving for revolution. All of the love, sex and booty music has led many African/Black males toward a path of self-destructive behavior. Only revolutionary rhythms that are independently controlled will ever open up an ocean of possibilities for success in a racist society. Take note that the same techniques used to enslave the mind are the same techniques needed to the free the mind.

Athletics

The greatest and most versatile athletes in the marketplace are those of African descent. In other words, melanin-dominant individuals excel in the sports where high levels of psychomotor skill development are on display. In football and basketball, the sports are dominated by unique athletes. For example, playing defensive back in the NFL requires great versatility to defend with speed, to tackle, to intercept, and to cover wide areas of the field against the other teams' best offensive players. Professionally, at the highest level of football, black athletes dominate the position. Over time, we have seen black athletes revolutionize the quarterback position, and this is no understatement to what we have seen from the roll call of Black quarterbacks. These superior talents do so much more than chuck a brown ball in the air. From Doug Williams winning a Super Bowl in an amazing fashion to the blazing speed of a Michael Vick to a Lamar Jackson, there is a certain rhythm on display. What power, however, do they have to speak up for their socially conscious fellow black quarterback (i.e., Colin Kapernick) who was white balled from the NFL. Our rhythm is off when we cannot collectively support an athlete who is speaking truth to impact change in our society.

The way Kapernick was treated by a white-controlled but black-dominated sport demonstrates that our rhythm is off. Although major professional athletes obtained multimillionaire status due to their talent, they are virtually highly paid slaves in a racist system controlled by white business owners (Rhoden, 2005). For basketball, we have heard debates on who is the greatest of all time. The debate usually centers around Michael Jordan and Lebron James, but we know basketball is a team sport, so the talent is spread out. Also, the generation compared makes a difference, so there are a host of other basketball players that are too numerous to mention that could be considered. However, the power and influence of what these athletes can do off the court is even more important. After basketball, both Jordan and James are influential businessmen, and their contribution to the community will be measured in their social deeds. Bouncing a ball is for boys but men with a conscious mission build. Rather than analyze the on-court accomplishments, let us follow their rhythm off the court. If they chose material gain over social responsibility, then what use are they to the community?

African/Black males have a certain, physique, attitude, drive, ability and talent that helps them excel in sports like football and basketball, and rhythm is one source of strength that can lead to a high performance. When an athlete is moving and grooving in a sport that requires complex psychomotor movements, it is rhythm that determines success. Opponents try to break your rhythm to stop you from excelling in your sport. If your rhythm cannot be broken, however, the athletic performance is unlimited. Simone Biles is the perfect example of the unlimited power of a melanin-dominant athlete. She has redefined gymnastics and her skill, determination and physique are not comparable. No other athlete in her sport has ever reached her accomplishments as a seven-time world champion. She is so good that some of her moves are named after her. Prior to Biles, there was Surya Bonaly in Figure skating. The establishment did not know what to do about her talents, so she was virtually ostracized. These amazing athletes are in their own rhythm and there a host of additional examples in track and field, tennis, hockey, and boxing where the performances are world class and unmatched by melanin-recessive athletes.

African/Black athletes are highly movement-oriented, and movement must be valued in their life at a young age. Besides providing them with a positive activity to participate in, they will be developing the necessary rhythm needed to succeed in many other aspects of life. Imagine leagues of any sport controlled and run by melanin-dominant athletes. With a revolutionary consciousness molded in their minds, it would throw off the rhythm of the former slave master/oppressor/white boss whose goal is to win the game and to never give up power. Due to the value placed on money, many African/Black athletes are too ignorant to really understand the game they are playing. Many are just gladiators and slaves to a system that thrives off of their rhythmic success. As the rhythmic pendulum swings, this exploitative scenario should not last forever. We have had our leagues before (Negro Baseball League, American Tennis Association), so one of these decades in the future, the out-of-whack distribution of wealth for a person who can run, jump, kick and throw a ball and punch will be controlled by those who play these games. True power will not arrive until education is instituted.

Education

The African Diaspora cannot begin to build a strong contingent of African/Black males without an educational system that is controlled by African/Black people. Since the former oppressors' control many of the sources of knowledge received by the black community, it can be said that miseducation is the rhythmic order of the day. Historically, people of African descent have made enormous contributions to humanity. There have been glorious as well as dismal periods in our history and this ebb and flow in history indicates that the black community cannot and will not remain in this state of miseducation forever. We need to break the shackles on our minds because mental liberation will be an unthinkable quest. Our former oppressors have us in a certain mesmerizing rhythm that deters reading and a quench for knowledge. To break the monopoly that the oppressor has on the mind of oppressed people, a different tune needs to be created that will lead to a revolutionary spirit.

Public and most private education today is essentially propaganda. The masses of people are being fed lies to ensure a docile and a passive society that is controlled by a few elite individuals. Through education, the most powerful way to conquer man is to capture his mind. In psychology, we are taught that there exists a conscious, an unconscious and a subconscious mind. The subconscious mind is where most of our internal mental processing takes place. It is believed that the subconscious mind makes up close to 95% of our mind. Everything that enters the subconscious is taken on face value. Since everything reaching the subconscious is a "truth" that directly effects our activities in the inner- and outer-worlds, this makes the subconscious mind especially important in terms of how we perceive our world and how we relate to one another (Nur, 2003). To continually call one another bitch, dawg and other derogatory names, it is apparent that we have been fed a language that cannot lead to mental upliftment.

To reach the subconscious mind, distraction and repetition are two forms of mind or mental programming that can be observed daily. Using sex and pornography for distraction can lead one to make poor judgements. Repeatedly portraying African/Black males as thugs can only lead to a self-destructive mental program. Knowing the techniques used to enslave the mind are the same used to free the mind. African/Black males have the power to change their situation in life and find a new rhythm.

What the conscious mind believes, the subconscious mind acts on. It works like programming a computer. Information is fed into a computer, and the computer acts on it. However, if the information you fed into the computer is wrong, it still acts on it. If you give yourself incorrect information or if others give you incorrect information, the memory banks of your subconscious mind do not correct the error but act on it. The sad fact is that misinformation occurs daily, and this dysfunctional programming is the source of the delinquent and destructive behavior we see in many African/Black males. The corporate elite who control the economics of our society thrive on the rhythmic dissemination of misinformation into our subconscious minds, and not just toward African/Black males.

Health/Fashion

Health and fashion are two important matters in which there is an imbalance in the lives of many African/Black males. For example, fast food and a lack of vegetables and fruits at a young age can create poor health as an adult. There needs to be a balance in the diet to create the proper rhythm for good health as an adult. With the high incidence of prostate cancer in African American men, preventive measures should occur early in life. African/Black males may have poor nutritional habits because they think it is uncool to eat fruits and vegetables. Beyond food, fashion statements also hinge on what is and what is not cool.

The only connection with fashions made is that African/Black males seem to be concerned with cultivating behinds rather than minds. In other words, the baggy pants showing underwear, men wearing earrings, and the many wild haircuts are all fashion issues that distract the attention of black males from thinking about revolution. Men, in general, are very visual. Therefore, styles, fads and fashions are a strong driving factor to pursue things. It is very interesting to note that many non-African/Black merchants make money off of the African American consumer. There is no rhythm to our spending habits, and our community suffers because we are seduced by material things. A decreased interest in expensive cars, exorbitant car accessories (e.g., rims) and jewelry will be a start to eliminate these distractions. African/Black males find value in impressing peers. No rhythm can ever be established if wealth is always spent on trying to impress someone else. Next, we will discuss interpersonal relationships to shed light on why fashions are so important to our youth.

Interpersonal Relationships

Due to our assimilation into western culture, we have strayed away from what is important to us. From an axiological perspective, we normally value people and not material things. However, with the focus on impressing people, material things have become the vehicle to be somebody.

Our rhythm has been thrown off and there is a need to return to what we value. We need not take on and model our behavior after Europeans (Baruti, 2002). It is unfortunate, but many African/Black youth act like real dogs with their tongues hanging out when they see a beautiful sister in their presence. The tongue swagger is really a sign of disrespect for woman. We need to regain a rhythm that will help sisters respect brothers who know how to treat women.

It is nice to admire the beauty of a woman, but too many African/Black males get distracted with issues pertaining to love, lust and sex. The root of this problem may be the stud mentality that was created during chattel slavery. Slavery destroyed the Black family structure and our rhythm has been out of sync ever since, and the media vilifying Black Males has had an impact. Today, it could be said that a Black woman could be more fearful walking down an unfamiliar street with a black male versus a white male. This despicable turn of events magnifies the self-hate and the dysfunctional interactions between the two genders that should be uniting and not fighting. If the family unit is to survive, then a rhythm must be sought.

Under normal conditions, male and female are supposed to combine as one. Neither males or females can procreate without the other. For mating purposes, there is a natural rhythm that engages males and females to exist as mates. There are numerous reasons why, but sometimes there is a desire for male-male or female-female relation-ships. Procreation cannot occur naturally with this unbalanced rhythm, and there are a host of social, cultural and biological reasons for homo-sexual relationships (Baruti, 2003). The point to make here is that heterosexual as well as homosexual relationships are all about finding a rhythm in a person's life.

Often times a person from one gender cannot relate to people of the opposite gender. Opposites are the natural rhythm of life for procreation, but some people have a driving inclination to go away from this natural way of life. The society we live in puts too much stress on relationships and the stress has an impact on how people interact. Women are primarily viewed by many men as sex objects and many women seek men for material gain. The core of these types of relationships are rotten, and the natural rhythm of interpersonal relationships are ruined.

Many young black males are visually oriented, so they are distracted by the physical appearance of women. They must be taught to not view women as sex objects and to understand the power and influence of Western culture on their mind set (Baruti, 2002). African-centered scholars must bring a new generation of Black males into a better appreciation for their African culture. In fact, a more controversial topic that deserves further attention is the role of man sharing in a society that has too many women and not enough eligible men. It is only controversial because we normally think of monogamous relationships as the natural rhythm to life. In many African cultures, polygamy was the social order of the day that kept a society intact. If we think about rhythm, the thrust is not to create a generation of "players," but to make young African/Black males understand the importance of treating all women like the mother who bore them. There is more to establishing and maintaining a healthy relationship than sex, and it is time to evaluate the dysfunctional relationships that currently exist as well as the high divorce rate. When people become mature, they can begin to understand the value of supporting one another in ways that create a rhythm. Big Brother and Big Sister programs as well as Male and Female mentoring programs shed light on the fact there are too many families with a missing father or mother. If we understood the unique balance of man sharing or polygamy and if we taught it to the next generation of African/Black youth, it might make them more mature and help them to deal with the current gender imbalance in our society. We must decrease the lust for a woman's body and increase the thrust for a nation-building mentality.

Marriage

A discussion on finding the right rhythm in relationships can lead to secure marriages. A full- length discussion on marriage is far too detailed for what can be provided here, but we must be aware that many African-centered cultures were and still are matrilineal (Diop, 1990). With the mother or female as the center of the family, there is a better appreciation for women. In the past and in some modern African societies, polygamy had been a valuable social order (Amen, 1992).

73

The danger today is to think polygamy is all about having a harem of free lovers. It is crucial to understand that polygamy is not a "style of marriage" or "marital preference" as some western scholars have claimed. It is an answer to an age-old social problem in which there is an excess of women in comparison to men (Amen, 1992).

According to Ra Un Nefer Amen, polygamy in matriarchal or matrilineal societies was established and advocated by the women. The entire society benefitted from such relationships. In all matriarchal or matrilineal societies, polygamous unions are marriages no less endorsed and regulated as are monogamous marriages. Women in polygamy are wives no different than women in monogamous marriages. To create a social rhythm, they have one mate to whom they have as full a commitment as does a woman in a monogamous marriage. Unlike the majority of sexually active single women, they have one mate to whom they have as full a commitment as does a woman in a monogamous marriage.

Once we understand that it has been practiced for several thousand years to guarantee every woman a family, and the means of satisfying her needs without having to hide, and expose herself to danger, then we should applaud these societies for having the moral courage to face the issue, and attempt a rhythmic solution that addresses the gender imbalance in our society. Interestingly, the average duration of marriages in America is seven years. This creates a scenario in which one man can marry (and divorce) or sustain a relationship with three women over a period of twenty-one years (Amen, 1992). Only mature people could reflect on this reality and not the Music Television Video generation of people who only see women as sex objects and men as an opportunity to gain material wealth. To find our Rhythm Nation, we must keep thinking of ways to remedy the dysfunctional marriages that lead to divorce (Moore, 2011).

Conclusion

"Da drum call" and the pulsating rhythm reverberates to communicate socially relevant information. The struggles faced by many African/Black youth require serious attention to raise the next generation of warriors. We need to find a rhythm in which people can relate to one another in a respectable manner. In this chapter, an attempt was made to invoke the ancient principle of rhythm as a source of strength for African/Black males. Creating a Rhythm Nation means that there is a collective interest in finding solutions to the many problems that exist in the African community. Since young African/Black males are movement oriented and prone to innovative creations, rhythm should be used to bring order to our community.

If a society can manipulate the minds of their youth, then the next generation of people is effectively controlled. Distraction and repetition are two mental programming techniques that ensure mental enslavement. When a person's subconscious mind is negatively programmed, it can dramatically influence group behavior. Many African/Black males are seduced by material gain and sexual conquest. Athletics, entertainment and fashion are key factors in this complex game of seduction. To counterattack these negative influences, a rhythmic balance must be found in health-related matters. In addition, proper education for revolution must be sought. True freedom will only occur when we are able to have respectable interpersonal relationships and lasting marital arrangements to build a Rhythm Nation.

History shows that when people do not take responsibility for their lives, there are those who will take it for them. The future builders of this Rhythm Nation must remember that once freedom is lost, it is much harder to regain, and the goal should not be a return to slavery. African/Black males are integral to the future of the African Diaspora, so get your mind off of the booty and focus on the finances. You must be economically self-sufficient and an independent thinker to successfully build a Rhythm Nation.

CHAPTER 7

PENIS POWER AND PIGMENTATION

Virility and no sterility.

Introduction

In a society that professes to be democratic in which all men/women are created equal, what major factors have driven the insane lynching of Black men, the opposition to miscegenation for legalized citizens, and the too frequent cases of police brutality against Black men? On a historical basis, what makes the ugliness come out in fragile people where there is a need to be xenophobic and vile against another? Even more shocking, the attacks and the laws against melanin-dominant people are against people who are deemed "inferior." If people of a darker hue were so inferior, there should be no laws needed to hinder their progress. Inferior people should not be able to compete anyway, so there are other motivating factors to consider. One common denominator in these negative situations is linked to the virility of melanin-dominant men.

Naked and Exposed

To emphasize this point, without money or material or a created job to distinguish one person from another, it is the nakedness of who you are that defines you. From an evolutionary perspective, Homo sapiens were born naked and unclothed in the harsh reality of nature. The human body adapted for outdoor living. This virility and natural power can be neutralized, however, when money, material and a job are introduced into the scheme. To be able to neutralize and negate is to have power. Imagine the fragile mind that is created when one must recognize their inadequacies when exposed and naked. To compensate for this lack, the person feeling inferior will use defense mechanisms to defend themselves. They may use projection and accuse the opposition of being the problem. Reaction formation may occur in which something that was bad is now good. The fragile person may also deny that they are feeling inadequate. The entire human experience is manifested on an intellectual, moral, physical, and spiritual level. No one area of the human experience should be more emphasized than the other if you are to be a whole person. In a civilized world, material items or material wealth should be the very last entity that defines who you are.

There are a multitude of psychological twists in the mind of people who feel inadequate around others, but our concern in this chapter is with the power of pigmentation, especially in the penis. Of course, the reaction from most melanin-recessive people would be that they do not feel inadequate to people of color. This reply is on a conscious level because they see themselves above melanin-dominant people in numerous categories. Subconsciously, however, there is a fear of "blackness." According to Frances Cress Welsing (1991), "The Cress Theory of Color Confrontation states that the whites or color-deficient Europeans responded psychologically, with a profound sense of numerical inadequacy and color inferiority, in their confrontations with the majority of the world's people – all of whom possessed varying degrees of color-producing capacity….As might be anticipated in terms of modern psychological theories, whites defensively developed an uncontrollable sense of hostility and aggression." These events are happening on a global scale.

Penis Psychosis

A social commentary is provided because the tragic history of Black men being wrongfully accused of assaulting, insulting or interacting with white women is all about sex. The concomitant psychological frenzy that develops in a white male fragile state of mind from the thought of such a relation or assault creates a psychopathic culture. Since the creation of the USA, many White men have exhibited racism and aggression toward melanin-dominant men and their so-called inferior communities (Jaspin, 2007; Dray, 2003; Ginzburg, 1988; Carr, 2007). These communities were only called inferior by the people oppressing them because it was about control. By controlling access to resources to survive in order to be independent and self-sufficient, many black communities were attacked by mobs and never permitted to be on a level playing field.

78

For a mob-like mentality to form there must be a certain animus driving this psychotic behavior; that animus is caught up in the loins of melanin-dominant men. If many of the lynching events were over the accusation of a white woman being assaulted, there was often a subsequent demonic lust to mutilate the genital region of the melanin-dominant male. Even if it was not for the false accusation of assaulting a white woman, there was a satanic desire to cut off the penis, with children present. The books, postcards and documentation are very revealing and disturbing as they put the white American psyche on full display (Allen, Als, Lewis, and Litwack, 2003). Call it insane, irrational or an aberration, but we know it is not normal. Deep in the thoughts of Freud's theory about penis envy, on a white cultural level, lynching was a penis psychosis.

Besides lynching people of color, there was another lethal means to neutralize this natural virility. The manufacture of the gun and laws implemented provided the ultimate weapon to effectively erase your personal inadequacies. With the projectiles from the gun in the form of bullets, it became a parsimonious solution to erase what is feared. Welsing (1991) emphasized this in her subject matter on the gun as a symbol. The fact that the projectile from a penis (i.e., sperm) can eliminate whiteness is a subconscious fear that has existed since the first encounters in written history. We can speculate on prehistoric times, but we know today that many western and Eurocentric cultures with a history of being a colonial power or an oppressor toward people of color, the gun was effective to stop the onslaught of the potential genetic annihilation of melanin-recessive people.

Along with the gun, the legal system and the injustice in the courts highlights the intense fear of melanin-dominant virility. Try and search for a better explanation for why innocent black citizens are being murdered by law enforcement officers or for the crazed support of gun activists to bear arms. Despite the many mass killings with guns in our "civilized" society (Pane, 2019), there is a lack of leadership to change the laws to protect the common citizen. We all live in danger solely because the innate fear of many white people who have committed massive violence on a global scale. It is as if the gun is needed to combat the fictional resurrection against this past injustice. We are all forced to live in a psychologically distraught world due to the crazed influence of psychopathic thinking (Wright, 1984).

Dark and Lovely

Let us discuss science. Did you ever contemplate why the penis is much darker than other body parts that are more readily exposed to the sun? Unless you are a nudist today, your penis is never exposed to direct sunlight. Whether circumcised or not, the shaft of the penis contains melanin and the genital region is dark for many reasons. First, it may be for protection of this sensitive area that contains the genetic material to procreate. Since ultraviolet radiation from the sun can kill cells, the body's natural response has been to provide external protection with extra-pigmentation. Secondly, the importance of mating must be considered. The added color may be attractive for the opposite species (i.e., male-female relations) to increase and enhance the success rate to reproduce. Similar to lower animals, the better chance to mate and carry on genes with an inviting reproductive organ. Although less significant in males, we can say the dark area around the nipples of males and females have extra pigment for cosmetic reasons. It can provide excitement to reproduce.

Thirdly, we can expound on the theme of this book. Melanin provides extra-sensory perception and it promotes properly functioning organs, such as the penis. The melanin heightens the functioning of the organ. The friction, the heat, the erection and the stamina associated with sex are all dependent on melanin-dominant physiological mechanisms. For example, it is interesting that the major factor attributing to the success of drugs used to treat erectile dysfunction is blood flow. Enhanced blood flow to the corpus cavernosum in the penis can enlarge the organ for an erection. Oddly, there are medications that can darken the skin without sunlight and these compounds can treat erectile dysfunction. Actually, the erections are a side effect from these melanin enhancing drugs that have now been marketed for the sex industry. In *Pigment Power*, we are showing the connection between scientific investigations and highly melanized systems.

Nitric Oxide and Sex

The commonly marketed drug called Viagra was initially used to regulate blood pressure. A side effect from the medication was an erection in male subjects. This led to the massive marketing of new drugs (e.g., Cialis) that manipulate blood flow. There are many organs that require enough blood flow, so we cannot be naïve that the penis is the only organ system affected by these drugs. Over the next few years, scientists may uncover a whole host of organ problems and illness associated with prolonged use of these synthetic drugs.

Current research has investigated the effects of the gas called nitric oxide because it can increase blood flow to the corpus cavernosum in the penis. Nitric oxide is a simple gas made up of one nitrogen and one oxygen molecule, and it makes the blood vessels more elastic. It is produced naturally in the human body, and it functions as a neurotransmitter in the brain. In the grand scheme of the field of neuroscience, it is a relatively new neurotransmitters that affects a host of critical factors ranging from cardiovascular health, improved immunity, brain boosting properties, sleep enhancement and increased sexual performance.

I have presented on ethnic weapons in the past to educate the reading audience that scientists can go to the lab and specifically target certain genetic populations. This is diabolical, but it is understood in the context of population control and corporate manipulations. Henceforth, controlling the market economy can happen over extended periods of time without the person realizing it. When children are watching television and they get bombarded with the "erection" commercials and the constant messages about drugs for sexual enhancement, we can see a generation being led astray. Not everyone needs these medications, but it is all about marketing, corporate control, and manipulation. It is a perfect storm for a sex crazed nation.

The speculation here is that the origin of compounds related to Viagra began in the homosexual culture during the 1960s. In her dynamic book, *AIDS, Opium, Diamonds, and Empire*, Banks (2010) reveals that amyl nitrite was developed to be used as a vasodilator for people who had a heart condition known as angina. It was a controlled substance requiring a physician's prescription up until 1960. Alkyl

81

nitrites are a source of nitric oxide, which signals for relaxation of the involuntary muscles. Banks does not suggest this, but the decades of observing the impact on the sex industry was lucrative. According to Banks, however, from 1961 to 1969, a subset of the homosexual community began using these nitrite related drugs as a sexual doping agent because of a transient high and anal sphincter relaxation. These drugs were a $50 million dollar a year money maker in the 1970s, and we fast-forward to more nitrite-related drugs almost 50 years later (i.e., Viagra and Cialis). The bottom line, sex can kill a black man at the end of a noose for dealing with a white woman, but sex sells for the corporate-controlling white male who subconsciously expresses a penis psychosis.

Melanocyte-Stimulating Hormone and Synthetic Replicas

On a natural level in a healthy male, melanocyte stimulating hormone (MSH) can enhance sexual desire. With research conducted in the laboratory to study MSH in humans, we can conclude that sexual desire and erections are influenced by this natural hormone that may exhibit different levels in ethnic groups. Considering that humans have a different hue is clear evidence that some people are genetically predisposed for these chemicals to work differently in the hueman experience. The darkness of the penis and the response to circulating chemicals like MSH can stimulate melanocytes and melanin and trigger factors that can heighten the sexual experience. Scientists have now exploited this natural system with the creation of drugs for melanin-recessive people to tan. The creation of these products is the scientific evidence supporting pigment power in the penis. Unlike Viagra and other related medications, drugs directly affecting MSH do not act upon the vascular system. Instead, these pigment potent drugs directly increase sexual desire via the nervous system. Without a doubt, both heads are connected, and some people get caught up thinking with the wrong head.

Just like the serendipitous finding that blood pressure medicines like Viagra can cause erections as a known side effect, drugs used to enhance pigmentation have revealed similar side effects. Companies such as Palatin Technologies, for example, have been in hot pursuit of FDA approval to market drugs that enhance sexual desire. Melanotan I and Melanotan II have been studied for years as a drug to tan the skin without ultraviolet light. In the Science of Melanin, I wrote 25 years ago that this market would be on the rise (Moore, 1995).

Melanotan II is a synthetic analogue of the peptide hormone alpha-melanocyte stimulating hormone. It was purported to be an effective "tanning drug" without the use of sunlight. The side effects, however, include uneven pigmentation, darkening or enlargement of existing moles and erections in males. Interestingly, it was first under investigation as a candidate for female sexual dysfunction, but clinical trials have been controversial. Unlicensed and untested products may be on the internet, but as of 2019, FDA approval is still being sought to market these drugs for commercial development.

When you search the internet and look for the product, it is sold as an injectable agent. Only a doctor can prescribe the drug, so consumers must be aware of what is on the internet for sale when the product is not regulated. The drug called Bremelanotide is the drug of choice. It is also called PT-141, and it is a peptide that was previously marketed as the female version of Viagra. It was first investigated for female hypoactive desire disorder, but a woman does not have a penis, so the effects are more than penile erections. Interestingly, the synthetic peptide known as PT-141 binds to melanocortin 4 and melanocortin 1 receptors. This is significant because those melanocortin receptors are complex and they are found in many regions of the body (Cone, 2000).

Melatonin Rise

In a healthy male, the penis is activated for an erection in the early morning hours and this is the natural effect from a circadian rhythmic response. Under normal circumstances, melatonin is released from the pineal gland at night and the pattern may extend from 11 PM to 7 AM. Serotonin is a precursor molecule released from the pineal gland during the day and serotonin can be converted into melatonin. Melatonin is a hormone that retracts the granules of melanin in the melanocyte to allow more light to penetrate the skin during the night-time hours. There is no sun shining at night, so melatonin is playing a role to lighten skin and allow more light to impact the body.

The penis is heavily pigmented, and it responds to melanin activation. With both melatonin and melanocortin receptors in the skin of the penis, the early morning erection can relate to the nighttime rise in melatonin. In nonhuman mammals, melatonin suppresses gonadal activity. With the shorter days and longer nights during the winter months, the sun does not shine as long. As a result, the extra melatonin circulates to suppress gonadal activity. It is not prudent to have babies in the winter because survival rates would be low for nonhuman mammals in the wild. Humans, in contrast, can reproduce any time of the year. There are cases of problems with puberty and hypogonadism in humans if the pineal is not functioning properly, but under normal circumstances, gonadal inactivation is not a major role for melatonin in humans.

Penis and Pigment Boosts

J.A.H. Diouck (2018) has written a superb book on a Melanin Guide to Spiritual Awakening. She is a master nutritionist and has formulated food elements that can enhance a melanin-dominant system. Her book will keep the reader healthy, fit and conscious. Other researchers have addressed diets for melanin-rich people (Meningall, 2008) and specific nutritional needs for melanin-dominant people (Afrika, 1989; 2000). In terms of sexual health, a colleague and expert chemist known as Bruce Ferguson and I were in the process of building a company to market Ashwaganda as an immunostimulatory compound with effects on sexual performance. However, on a simple level to combine what has been read in this chapter, any food items that can stimulate nitric oxide or function as a vasodilator are critical. Beets are a good example of a potent food item that contains high levels of pigments, vitamins and nitric oxide.

Not only do beets contain a sufficient level of nitric oxide, they also contain methyl groups, which are chemicals that can greatly lower estrogen levels in the body. Lowering estrogen in males can allow testosterone to flow freely through the body to increase weight loss and muscle growth. Certainly, we know testosterone levels affect sexual desire, and the pigment in beets provides the power to boost this steroid hormone. In addition, this is why beets are often used by athletes and other people who perform demanding activities since beets can help their muscles develop and remain strong. Therefore, think about the penis as a muscle that can stay strong.

Melanin-dominant people have the necessary chemicals that can naturally boost melanin and enhance sexual performance. There is a long history of drugs and various chemicals that have been investigated to explore the stimulatory as well as side effects associated with sexual activity (Lieberman, 1988), and what has been presented in this chapter has not been emphasized in the literature. If a person lacks the melanin possessing compounds that are being studied, it can be difficult for that person to extrapolate the subjective feelings. Natural elements that have been used by ancient cultures for centuries are healthier than the synthetic products created in the lab. Beets were suggested as a powerful natural substance but investigate what ancient cultures used. Horny goat weed is known to be an aphrodisiac, and Maca root has been used for libido. Know thyself and control thyself.

CHAPTER 8

MELANIN AND CELLULAR ENHANCEMENT

Got Melanin?

In Chapter 7, information on the male genitalia as a dark and lovely organ was presented to emphasize the importance of this anatomical region for procreation. The deep science was avoided so the reader would better grasp the politics associated with the power of the penis as a melanin-dominant organ when compared to other external body parts. In this chapter we will go further into the science behind the way melanin and chemicals associated with pigmentation can affect a host of cellular enhancing mechanisms.

The smooth tone of hue on the outside of the body is primarily melanin which is synthesized by melanocytes in the basal layer of the epidermis. Melanin oozes out of the dendritic arms of the cell to provide an even pigment. When you have a genetic abnormality like albinism, vitiligo or piebaldism, melanin is either absent or spread unevenly. For instance, albinism is due to a genetic abnormality for the enzyme tyrosinase, vitiligo is a disease which causes the loss of skin color in blotches, and piebaldism is a rare autosomal dominant disorder of melanocyte development.

During cell division, there is evidence that not all melanin is distributed symmetrically (Joly-Tonetti, Wibawa, Bell and Tobin, 2018). During mitosis, a mysterious presence of melanin can be differentiated into keratinocytes. The authors conclude that steady-state epidermis pigmentation may involve much less redox-sensitive melanogenesis than previously thought, and at least some pre-made melanin may be available for reuse.

In other words, melanin is not a waste product, and the abundance of melanin in the epidermal unit is histologically complex. As we have noted, with extra pigmentation in the male penis and scrotum, this research by Joly-Tonetti et. al. (2018) provides scientific evidence for the inheritance of asymmetric distribution patterns. Again, the penis should not be darker than your forearm, but an asymmetric pattern of melanin distribution may explain the variety of tones in a melanin-dominant body.

Even in the foreskins of black and white subjects there is a highlighted difference in melanin activity (Iwata, Corn, Iwata, Everett and Fuller, 1990a). In this study, tyrosinase activity was assayed in black and white human foreskin samples by measuring both the hydroxylation of tyrosine to dopa (tyrosine hydroxylase activity) and the conversion of [14C] tyrosine to [14C] melanin (melanin synthesis assay). Tyrosinase activity in black foreskin homogenates averaged almost three times that in white skin samples (33.8 pmols 3H2O/h/mg skin in black and 12.71 pmols 3H2O/h/mg in white skin), although considerable overlap in activities existed among the two groups.

Normally, tyrosine hydroxylase activity is linked to neurotransmitter synthesis, but data from this study on human foreskin revealed that tyrosine hydroxylase activity is tightly coupled to melanin synthesis. To express the differences in melanin-dominant samples, tyrosinase activity determined by either assay method generally correlated with skin melanin content. Iwata et al. (1990a) concluded that immunotitration experiments suggested that the difference in tyrosinase activities between white and black skin may be due, not only to different amounts of enzyme present in the melanocytes, but also possibly to differences in the catalytic activities of the enzyme found in melanocytes of black and white skin.

The same team of researchers from Oklahoma Health Sciences Center (Iwata, Iwata, Everett and Fuller, 1990a) also investigated human foreskin organ cultures in response to hormonal stimulation. Both black and white human foreskins can be maintained in organ culture for at least 1 week with no change in the tissue structure or cell viability as determined by histochemical staining and by dopa reaction staining. Tyrosinase activity in both black and white human foreskin cultures decay markedly during the first two days of culture to a new steady state level which remains stable throughout the culture period. Both black and white foreskin cultures consistently demonstrate 2- to 10-fold increase in tyrosinase activity when treated with theophylline (1nM).

The connection with theophylline is interesting because chocolate is made from cocoa beans, which contain theobromine. Theophylline and theobromine are mild bronchodilators and they affect mood. It is speculated here that the presence of these substances in chocolate can create arousing effects at the level of the skin to have an influence on the euphoria some people feel who consume chocolate.

In addition to theophylline, Iwata et al. (1990b) found that c-AMP and alpha-MSH markedly stimulated tyrosinase activity in some cultures. They conclude that the hormone-responsive organ culture can be utilized to characterize the molecular processes responsible for the regulation of tyrosinase and pigmentation in human skin. To reiterate the link with chocolate, which is primarily found in cocoa beans on the African continent for centuries, reflect on what was explained about the dark and lovely expression given to the penis in Chapter 7. There was a 2- to 10-fold increase in enzyme activity when theophylline was administered to foreskin.

Research on foreskins can be considered political and controversial in the public realm for a host of reasons. First, foreskins are being used as products as an active ingredient in facials for women. For example, the Oprah-touted skin cream from SkinMedica uses foreskin fibroblasts that are used to grow and cultivate new cells. Secondly, since research has been ongoing for decades, we cannot overlook the warning people like Dick Gregory has shared about the Atlanta Child Murders from the 1970s. He stated in his public appearances that many of the missing children that were found had their genitalia mutilated and this was linked to the big business of interferon research. Gregory's speculation was that the highest content of interferon was found in the head of the penis in melanin-dominant people. I cannot fully confirm this as factual and if the penis was mutilated, but it seems strange that we really have no trusting explanation for all of the missing children decades after the tragic events. Wayne Williams has been accused of the murders of adults during this time period, but to link him to all of the missing children is highly suspect. Decades beyond the Atlanta Child Murders, we know there is the illegal trafficking of organs, so anyone gone missing could be a possible candidate as an involuntary organ donor.

The timing of the research on interferon is very intriguing. The epidermal melanocyte system in newborn human skin was studied by Glimcher, Kostick and Szabo in 1973. Just before this study was an article written in German called Morphology of the Epidermal Unit by Frenk and Schellhorn (1969). It is difficult to locate this article in German, and who knows how much research has never been written for publication. We know, however, some German scientists were studying how to "get black," and Firpo Carr (2003) calls them "Melanin Marauders."

Some of the early German experimentation on black skin was from the 1940s to seemingly learn how whites could better adapt to tropical environments. To take back what Colonial German lost in Africa, Nazis had to learn how to become acclimated to the stifling, oppressive African sun, so according to Carr (2003), they literally squeezed the life out of Black Prisoners of War in order to discover the high level of tolerance to the sun that was intrinsic to black skin.

With the CDC located in Atlanta and many German scientists coming to America after WWII, the use of their scientific technology could have been used for more diabolical means in the USA. It is believed that these earlier studies led up to the billion-dollar cancer industry to study interferon and other immune-modulating factors such as interleukin. As more links are made, the research on chemical biological warfare and agents to shut down the immune system (Somani and Romano, 2001) were reported in the late 1960 before the boom of the AIDS epidemic in the 1970s and 1980s.

In relation to melanin, Zhou, Ling and Ping (2016) found that interferon is an important cytokine which can be secreted by keratinocytes or macrophages induced by UVB irradiation in skin. Mammalian skin cells have the capability to produce and metabolize serotonin whose cutaneous effects are mediated by the interactions with serotonin receptors. Treatment with serotonin resulted in a dose-dependent increase of tyrosinase activity and melanin contents in normal human foreskin-derived epidermal melanocytes while with interferon a decreased effect resulted. Whether treated with serotonin for 5 days or 12 days, the pigmentation level neither recovered after displacing the interferon containing medium. The suppression of interferon on serotonin-induced melanogenesis further suggests the negative role of interferon in inflammation-associated pigmentary changes.

The connection between cancer fighting agents such as interferons (Roth, Morant and Alberto, 1999) and interleukins (Zhou, Shang et al., 2013) and the epidermal melanin unit demonstrates the effect of melanin on cellular enhancement. Normal human skin relies on melanocytes to provide photoprotection and thermoregulation by producing melanin, and the growth and behavior of melanocytes are controlled by many factors. Zhou et al. study was conducted to investigate the effects of interleukin-18 (IL-18) on melanocytes and to elucidate the underlying mechanisms. It was shown that IL-18 increased the tyrosine activity and melanin content in normal human foreskin-derived epidermal melanocytes. These results in vitro showed the accommodation of IL-18 in melanocyte growth. Therefore, the authors suggested an important regulating action of IL-18 to melanogenesis and cell growth ability of skin melanocytes.

We briefly discussed melatonin in Chapter 7 to associate a rise in melatonin with human penile erections. Although it was an animal study, Olmez and Kurcer (2003) studied the effects of melatonin on alpha-adrenergic-induced contractions caused by electrical field stimulation or the alpha(1)-adrenoceptor agonist phenylephrine (Phe) on isolated rat penile bulb. The effect of melatonin on Phe-induced contractions was completely reversed by treatment with tetrodotoxin (TTX) and Vasoactive intestinal peptide (VIP) antagonist. Since alpha-adrenoceptor blocking agents are known to interfere with detumescence of the erect penis, serum levels or the administration of melatonin may affect erectile function. The effect of melatonin may be the result of its allosteric interaction with the presynaptic receptors on VIPergic neurons.

Lastly on the topic of melatonin and penile function, there was a genetic explanation in Fragile-X syndrome (O'Hare et al., 1986). Melatonin profiles were studied in five males with cytogenetic features of the Fragile-X syndrome including megalo-orchidia and macrogenitosomia. In comparison with age matched normal controls, the fragile-X group showed lower melatonin values and a significant impairment of the nocturnal rise of melatonin from the pineal gland. Melatonin deficiency may thus be responsible for some of the phenotypic features such as enlarged genitalia.

Melanocortins

The final topic on melanin and cellular enhancement to emphasize in this chapter will be the receptors in the body that respond to chemicals that can induce pigmentation. The process relates to melanocortins or melanotropins. Just like melatonin, there are melanocortin receptors responsive to these chemicals to control pigmentation, but the receptors for melatonin are different than the receptors for these ubiquitous peptides. The end result is that melanin is regulated by these endogenous chemicals with no need for the sun as an external agent to affect external pigmentation.

From the beginning, there is a large protein molecule called proopiomelanocortin (POMC) that contains numerous small chained amino acid sequences called peptides that can be cleaved from the larger molecule (Figure 9).

PRO-OPIOMELANOCORTIN

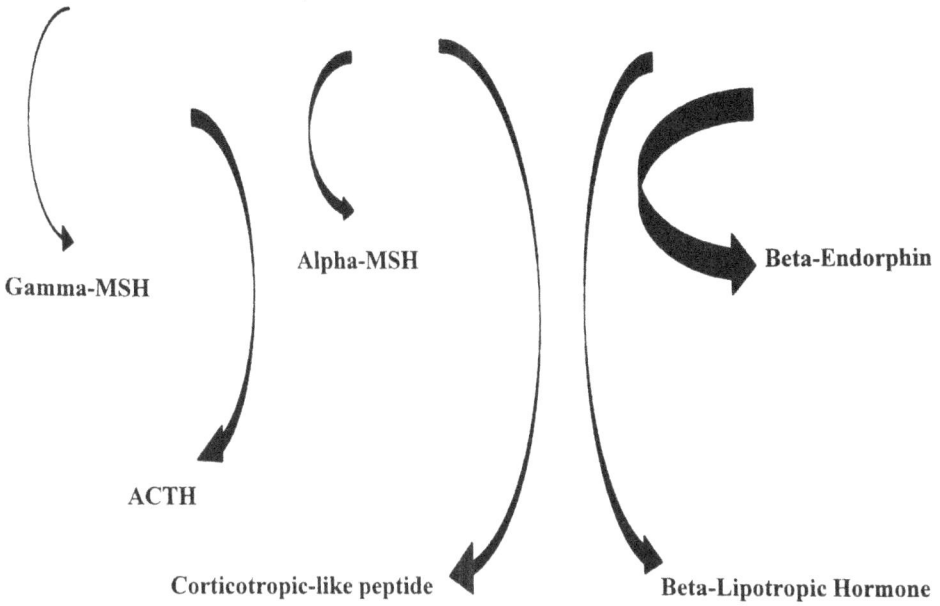

Gamma-MSH

Alpha-MSH

Beta-Endorphin

ACTH

Corticotropic-like peptide

Beta-Lipotropic Hormone

Sample Image of Five Melanocortin Receptor Subtypes

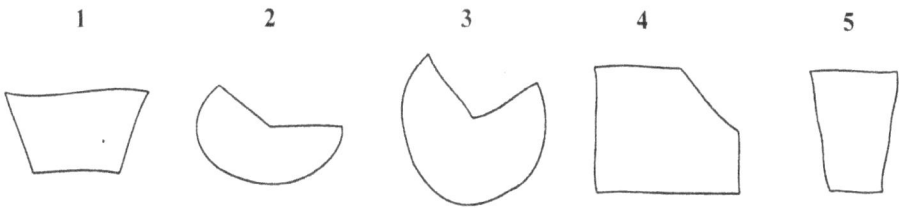

1   2   3   4   5

Fig. 9 - Pro-opiomelanocortin is the precursor or "Mother" molecule
which contains within its primary structure the sequences of
six melanotropic peptides: α-MSH, β-MSH, γ-MSH, ACTH,
CLIP and β-LPH. These peptides can interact with five different
types of melanocortin receptors.

Alpha-MSH, Beta-MSH, Gamma-MSH, Delta-MSH and ACTH are specific melanocortin peptides. The structural feature characterizing all MSH sequences and that of ACTH is the core tetrapeptide His-Phe-Arg-Trp, which is crucial for the interaction with the receptors of these peptides. They have been identified in the pituitary gland, the brain and various peripheral tissues of all classes of vertebrates either by bioassay, radioimmunoassay, immunocytochemistry, *in situ* hybridization or PCR.

The effects of melanocortin peptides on brain development, the developing motor system as well as on the regeneration of the PNS or the CNS is well documented (see Cone, 2000). The occurrence of POMC-producing cells is not confined to the pituitary gland and the BRAIN; it is well documented that POMC-mRNA and POMC-derived peptides are also synthesized in various peripheral tissues such as the SKIN, testes, ovaries, placenta, adrenal medulla, gastrointestinal tract, cells of the immune system, and in tumor cells.

The development of melanocortin-derived sequences for the treatment of mild forms of dementia, actively pursued for many years, appears to have been discontinued, despite the findings that peripherally and centrally applied melanocortin analogs and fragments positively influence cholinergic, adrenergic, and dopaminergic neurotransmission. Alpha-MSH can effectively increase the firing rate of dopaminergic neurons, and dopamine, as a neurotransmitter, is highly associated with euphoria along with the addictive nature of drugs.

Drugs require specific receptors to have an effect. For melanocortins, there have been five types of receptors (1 through 5) discovered, and they have a wide range effects on human behavior (Cone, 2000). The results from some studies suggest that alterations in the central melanocortin system may contribute to the clinical or behavioral manifestations of certain psychoactive drugs. Since we know melanin is expressed differently in humans, this can explain ethnic specific differences in both recreational and prescriptive drugs that can lead to addiction. In this chapter, we have shown a plethora of scientific mechanisms that help melanin to enhance life. These chemicals and the associated receptors play a key role in stimulating melanin throughout the body, and the presence of the same chemicals in the brain and the skin demonstrates a strong connection to boost the power of pigment functioning throughout the body. Even without the sun, there are natural elements to keep the body synchronized for cellular enhancement. In sum, POMC, MSH and the various types of melanocortin receptors have a powerful effect on the human system.

CHAPTER 9

MELANIN PROTECTION, STRESS AND AGING

Black don't crack.

Pigment Power Protection

A melanin-dominant biological system is enriched with the natural antioxidant capabilities to keep the body healthy. Pollution exists all around us, so the human body must be steadfast and ready to fight during every living moment. The pigment power of melanin provides the protection against the daily onslaught inside and outside of the body. Even in death, the presence of melanin still exists and is an indestructible life saver.

When the topic of melanin is related to stress and aging, the first inclination is that the protective properties of melanin can keep cells in a healthy state of functioning. For example, melanin is a stable carbon-based molecule, and it can go from a stable state to an excited unstable state when exposed to stress. The stress we are referring to can be the nonspecific response of the body to an outside influence that can affect the integrity of the cell. Illicit drugs, pharmaceutical agents, negative thoughts, obscene behaviors and a host of other factors can make melanin unstable and attribute to disease. The diseased state can often be a result from DNA genetic damage.

Skin pigmentation is important not only from cosmetic and psychological points of view, but more importantly because of its implications for the risk of all type of skin cancers and aging. Caucasian and Asian skins have highly similar gene expression patterns that differed significantly from the pattern of African skin (Yin et al., 2014). Even beyond the skin, Dubey and Roulin (2014) reviewed the literature indicating that internal melanin protects against parasites, pollutants, low temperature, oxidative stress, hypoxemia and UV light, and is involved in the development and function of organs.

In addition, studies (Mackintosh, 2001) on the antimicrobial properties of melanocytes, melanosomes and melanin have been reported to be an evolutionary advance. Evidence is presented that melanization of skin and other tissues form an important component of the innate immune defense system. A major function of melanocytes, melanosomes and melanin in skin is to inhibit the proliferation of bacterial, fungal and other parasitic infections of the dermis and epidermis. According to Mackintosh (2001), this function can potentially explain: (a) the latitudinal gradient in melanization of human skin; (b) the fact that melanocyte and melanization patterns among different parts of the vertebrate body do not reflect exposure to radiation; and (c) provide a theoretical framework for recent empirical findings concerning the antimicrobial activity of melanocytes and melanosomes and their regulation by know mediators of inflammatory responses.

In previous literature first written by Moore (1995; 2004), the statement was made that BRAIN MELANIN IS GENETICALLY PROGRAMMED TO FUNCTION AT DIFFERENT CAPACITIES DEPENDING UPON A PERSON'S OVERALL CAPACITY TO PRODUCE MELANIN. Nearly two decades later, we have continued evidence supporting this view. In review of the evolutionary and biomedical consequences of internal melanins, Dubey and Roulin (2014) have shown that the amount of melanin deposited on the external body surface is correlated with the amount located inside the body. This finding raises the possibility that internal melanin plays more important physiological roles in dark than light-colored individuals. It is very critical to note that internal melanin and coloration may therefore not evolve independently. Figure 10 illustrates the presence of cells containing neuromelanin in the midbrain.

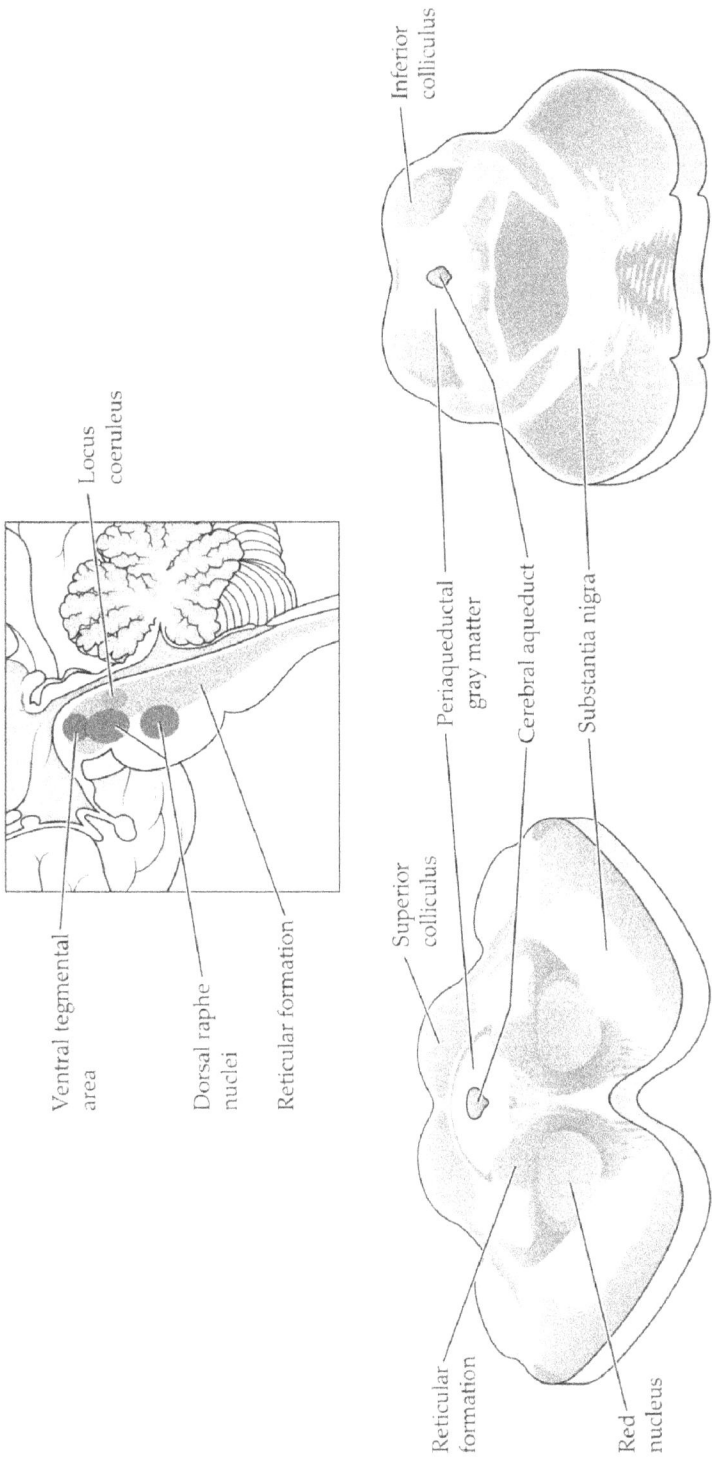

Fig. 10 – Sagittal and coronal section of the midbrain revealing pigmented neurons.

Locus coeruleus

Ventral tegmental area

Dorsal raphe nuclei

Reticular formation

Inferior colliculus

Superior colliculus

Periaqueductal gray matter

Cerebral aqueduct

Substantia nigra

Reticular formation

Red nucleus

According to Liala Afrika (2009), the negative unstable excited melanin state can cause toxic chemicals that may result in the destruction of melanin and cause disease. Carol Barnes (1988) had previously written on the harmful effect of toxic drugs on melanin centers within melanin-dominant bodies, so we know melanin can absorb and release these toxic elements in an uncontrolled manner. Both Afrika and Barnes emphasize that melanin attaches to synthetic chemicals as a neuroprotector. However, continuous consumption of the synthetic chemicals can over saturate the biological system, get into the blood stream, and harm every organ or cell with a high melanin content. Synthetic chemicals eaten get absorbed by melanin in the digestive system, then go to the brain and sex organs. The digestive system dumps synthetic chemicals from the intestines into organs such as the liver and kidneys, and we know disruption of these organs can wreak havoc on the human experience.

On the topic of food and drugs in the melanin-dominant body, Afrika (1989; 2000; 2009; 2015) as well as Whitaker and Fleming (2005) have written influential texts. The research from these scientists has alerted melanin-dominant people to be aware of what they are consuming. Therefore, let us consider the impact of genetically modified organisms in the next section.

The GMO Mystery

A current literature search on the topic of genetically modified organism (GMO) effects on black people revealed no results. Therefore, there is no reported research on GMOs and melanin. Although not related to melanin, a thorough book written by Lambrecht (2001) provides detailed information on how genetic engineering is changing what we eat, how we live, and how GMOs influence world politics. In fact, the exclusion of GMO labeled foods and the legal campaign to ensure companies are not required to label their products (Harmon, 2018) demonstrates that we should pay attention to why there is no research documented.

In the laboratory, we know there are ethnic-specific drug effects (e.g., blood pressure medicines). If drugs can be manipulated in the lab to influence different cellular groups, it is possible that GMOs can have ethnic-specific effects at the cellular level. This GMO mystery is problematic, and it sheds light on the global politics of food production and consumption. For example, pay attention to the food deserts that are forming throughout the USA in spaces occupied by melanin-dominant people (Torpy, 2019). Observe how the conglomerates like the company Monsanto have monopolized seeds and strangled both farmers and economies of color. Given this reality, we can assess that GMOs can be potentially hazardous to the human experience if not properly monitored. Controlling food production can be good when there is a benevolent mission to help people in need. However, most decisions made by big manufacturing companies are about making money and the health of people is not a primary concern.

In the food deserts are cheap dollar stores (e.g., Dollar Tree, Dollar General, Family Dollar) as a replacement for the vanishing healthy supermarkets (Torpy, 2019). The food in dollar stores is boxed, canned and frozen. The storage of this type of food requires chemicals and GMO products for a long shelf-life. With a concentration of food deserts, the health consequences are dire. Diabetes, heart attack, stroke, hypertension, cholesterol problems and cancer can be attributed to a poor diet. A poor diet in combination with food that has been genetically modified can wreak havoc and create chronic diseases that cannot be healed. In fact, this is a perfect plan for a capitalistic system because doctors cannot make money on healthy people. The use of GMOs is a perfect plan to usher in illness, and concomitantly, eliminate masses of people.

Money and profit reign over health and wellness, so we know that the consequences of GMO product use for the planet will continue to be debated. In fact, The National Bioengineered Food Disclosure Standard will require the disclosure of bioengineered foods by 2022. You should want to know what you are consuming, but companies have lobbied government officials, so the companies do not have to disclose that information. For a note, GMO produce in the market begins with an 8 for the four digit # on the label. If there is a five digit #, it should be organic.

Corn, soybean, potatoes, squash, papayas, apples, alfalfa and sugar beets have been approved by the USDA to be genetically modified. As a naïve population, we are in trouble because more than 90% of all soybean and corn grown in the USA is considered to be GMO. Therefore, food derivatives of soybean and corn products contain GMO ingredients.

Most GMO products are herbicide tolerant and resistant to infestation and disease. That means farmers can more liberally use herbicides and pesticides, and those toxins end up in our food. Studies indicate serious health risks may be associated with GMO consumption (including infertility, accelerated aging, and liver dysfunction). We know the melanin molecule absorbs chemicals and substances that are natural, artificial as well as genetically modified. Melanin may neutralize the chemical elements that are created from the consumption of GMOs, and genetically speaking, GMOs may change the entire physiological process.

We know our bodies have a multitude of mechanisms to fight disease and keep us healthy, especially during times we are unaware. Molecules that are created in the body from GMO consumption are probably undetected by current science. However, the body knows what is foreign and it will intelligently respond. Inflammation could be a response and the accumulation of toxic foreign elements can also accumulate in areas of the body in which melanin is serving as a protective agent against inflammation.

The increase in the number of neurodegenerative diseases is a serious implication that could be related to a genetically modified diet. Alzheimer's disease is a neurodegenerative disorder that is really responding like an inflammatory response.

Similarly, Parkinson's disease can be considered an inflammatory response that causes cells in the brain containing neuromelanin to die out (Figure 11).

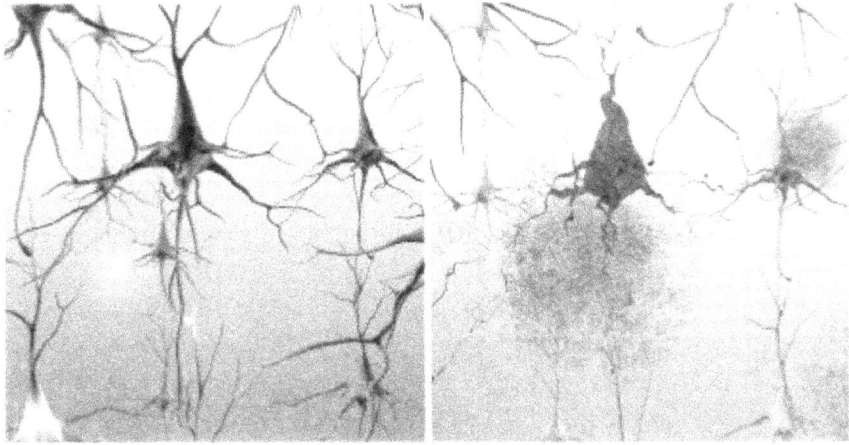

Fig. 11 - The neurons on the left are healthy and the neurons on the right reflect brain deterioration.

We must, therefore, take serious note of this fact because the unknown cause of many diseases might be GMO-related. GMO foods are made to resist bug infestation and to last longer on the shelf. Pesticides and/or herbicides can be added to make a six-week old food item, like fruit, look brand new and freshly picked.

In the long run, the food consumed may be creating by-products that can negatively impact physiological mechanisms. Although Alzheimer's and Parkinson's Disease can have a genetic origin, most of these neurodegenerative diseases can be idiopathic. Actually, the so-called unknown cause of these diseases might be the polluted food over decades of consumption. Next, we will look at the impact of experiencing decades long of psychological stress and how changes in the DNA can occur.

Stress, Aging and DNA

Physically, people of varying ethnicities respond to stress differently. It is not a matter of the cognitive difference we are referring to, but it is the difference in the cellular unit possessed by the person under stress. Since our discussion is on pigment power, we would like to compare and contrast the cellular differences between melanin-dominant and melanin-recessive people. Using disease models can reveal interesting results about the brain.

For instance, when explaining the DNA damage in pigment disorders, we can directly explore distinct differences in how the pigment cells are malfunctioning. There is a term called senescence in human skin, and it is recognized as a physiological and pathological change in aging and age-related diseases *in vivo* (Bellei and Picardo, 2019). This research team from Italy expands the definition of senescence to include the growth arrest caused by various cellular stresses, including DNA damage, inadequate mitochondria function, activated oncogene or tumor suppressor genes and oxidative stress.

Bellei and Picardo (2019) investigated melasma and vitiligo as two opposite acquired pathological conditions related to skin pigmentation and premature senescence. In both cases it was demonstrated that pathological dysfunctions are not restricted to melanocytes. Similar to physiological melanogenesis, dermal and epidermal cells contribute directly and indirectly to deregulate skin pigmentation as a result of complex intercellular communication. The difference may arise in melanocyte intrinsic differences and or in highly defined microenvironment peculiarities. This validates why there are differences between melanin-dominant and melanin-recessive people.

In Slovenia, Godic, Poljsak, Adamic and Dahmane (2014) reported on the role of antioxidants in skin cancer from exposure to reactive oxygen species (ROS) and oxidative stress. The only endogenous protection of our skin is by melanin and enzymatic antioxidants. Melanin is the first line of defense against DNA damage at the surface of the skin, but it cannot totally prevent skin damage. A second category of defense is the repair process, which removes the damaged biomolecules before they can accumulate and before their presence results in altered cell metabolism. As noted by Godic et al. (2014), additional ultraviolet protection includes avoidance of sun exposure, usage of sunscreens, protective clothes, and antioxidant supplements.

These European researchers (Poljsak and Dahmane, 2012) from Slovenia also investigated free radicals and extrinsic skin aging. Extrinsic skin damage develops due to several factors: ionizing radiation, severe physical and psychological stress, alcohol intake, poor nutrition, overeating, environmental pollution, and exposure to UV radiation. It is estimated that among all of these environmental factors, UV radiation contributes up to 80%. The primary mechanism by which UV radiation initiates molecular responses in human skin is via photochemical generation of ROS, mainly the formation of superoxide anion, hydrogen peroxide hydroxyl radical and singlet oxygen. From the authors, it is recognized that the only protection of our skin is in its endogenous protection (melanin and enzymatic antioxidants) and antioxidants we consume from the diet (vitamin A, C, E, etc.). With the intensity of the sun from the global climate crisis, it is imperative for melanin-recessive people to find interventions to prevent oxidative stress and to enhance DNA repair.

In Germany, there was an exploration into graying hair as a model to study topics related to pigment power. For example, Arck et al. (2006) studied oxidative stress and cell death of melanocytes in human hair follicles. Apoptosis refers to programmed cell death and a multitude of factors can destroy the integrity of the cell. Radiation, inflammation and/or psychoemotional stress can speed up the aging process, and graying is a prominent feature of aging. These authors found evidence for melanocyte apoptosis and increased oxidative stress in the pigmentary unit of graying hair follicles. Furthermore, there was a mitochondrial DNA-deletion associated with graying. The conclusion was that oxidative stress is high in hair follicle melanocytes and leads to their selective premature aging and apoptosis. Since melanin-recessive scientists are intrigued with slowing the aging process, this unique model system can be used to explore antiaging therapeutics to slow down or even stop this process.

A research team from Belgium (Van Neste and Tobin, 2004) using this similar melanocyte hair follicle model reported that graying is the result of reduced tyrosinase activity within hair bulbar melanocytes. By 40 years of age, there appears to be a genetically regulated exhaustion of the pigmentary potential of each individual hair follicle leading to the formation of gray and white hair.

In addition to this unique model to investigate the development of gray hair in humans, the exploration of cells cultured in the lab can be a useful model to study melanin protection and the aging process. In France, Cario-Andre et al. (1999), for example, had previously used a combination of cell culture techniques to study pigmentation and photoprotection in the epidermis. They used the air-liquid interface to grow differentiated keratinocytes, with the addition of 5% melanocytes in the seeding suspension. Therefore, it is not only melanocytes that are involved in the epidermis, keratinocytes have just as important of a role in photoprotection. Since these experiments can be studied in various skin types, the cell culture model allows the study of the physiology of the epidermal melanin unit as well as pathologic conditions.

It is interesting to see this extensive research from countries in which there is a minimal presence of melanin-dominant people. It is as if those who lack melanin are the most interested in the cellular mechanisms related to melanin functioning. Along with the human studies and the cell culture experiments, animal experiments can also demonstrate a connection between melanin, stress and DNA changes.

For example, Presse et al. (1992) studied melanin-concentration hormone (MCH) in a chronic footshock stress experiment in rodents. MCH is a neuropeptide that may be involved in the control of the hypothalamic-pituitary-adrenocortical axis and, more generally, of specific goal-oriented behaviors. In this study, chronic intermittent footshock stress causes a 58% or 29% decrease in messenger RNA content for MCH in the whole hypothalamus after 1- or 3-day regimen. These results provide evidence for a negative regulation of messenger MCH gene expression by stress and suggest a major role of stress hormones in a positive feedback control of MCH gene activity. Although this experiment is with animals, it helps to extrapolate the triggering effect of stress on systems which directly control pigmentation. Bad triggers can destroy DNA and the next section explores research making the link between stress and DNA damage.

Telomere Research and Melanin-Dominant People

Before we get into the science, reflect on the constant stress and trauma associated with melanin-dominant people being forcefully taken from their African homeland as prisoners by enslavers who exploited them and did not value them as humans. On a daily basis, we do not teach much about slavery, and the issue of reparations on a global level is virtually ignored. The historical amnesia is killing humanity. It is egregious to simply say, "get over it, that is the past." In the grand scheme of events, the enslavement experience was totally devastating to humanity for both the oppressor and the oppressed. For our topic, we can observe the genetic aberrations that have been passed down from generation to generation for the oppressed. Mind, body and spirit have been warped and the ability to imagine what humans really can be has been tainted by a melanin-recessive, materialistic reality of viewing the world. We must fight the influence of this inhumane negativity and return to knowing self to have any chance of understanding pigment power.

To reiterate, it is my strong belief that nothing in our recorded human history compares to the enslavement experience of melanin-dominant people. It was a war on a culture and the war has not ceased (Jones, 1993). Collectively, an entire group of people were kidnapped and extracted, beat and brutalized, raped and maimed, and left for dead in a foreign land (Anderson, 1995; Baruti, 2005; Clarke, 1996; Musta-keem, 2016). Not speaking your own language or eating your own food or having familiarity with your surroundings or being yanked away from loved ones and separated from your family is socially horrendous. It is not so obvious, however, what the impact might be on an epigenetic level, especially when the emphasis is, "just get over it, that is the past."

What else compares to this onslaught that has been recorded for nearly 400 years? Unless the mind has been warped via mind control, it should not be difficult to determine who is the real enemy. This is trauma on a monumental scale, and we can label this the MAAFA. Scholars like Sowande Mustakeem (2016) have shown impeccable research skills to investigate the horrors of enslavement, and these investigations are winning awards. Therefore, we can study the biological impact of the MAAFA. It is still an ongoing experience, and we can see it today in the escalation in police brutality, the unfair housing practices, the miseducation in the public schools, the discrimination on jobs, and the constant assaults on character. Scientists today have the tools to look deeper into what the MAAFA can do to the human body. On an epigenetic level, studying telomeres can provide answers to what is happening in the body.

A telomere is a region of repetitive nucleotide sequences at each end of a chromosome, which protects the end of the chromosome from deterioration or fusion with neighboring chromosomes (Figure 12).

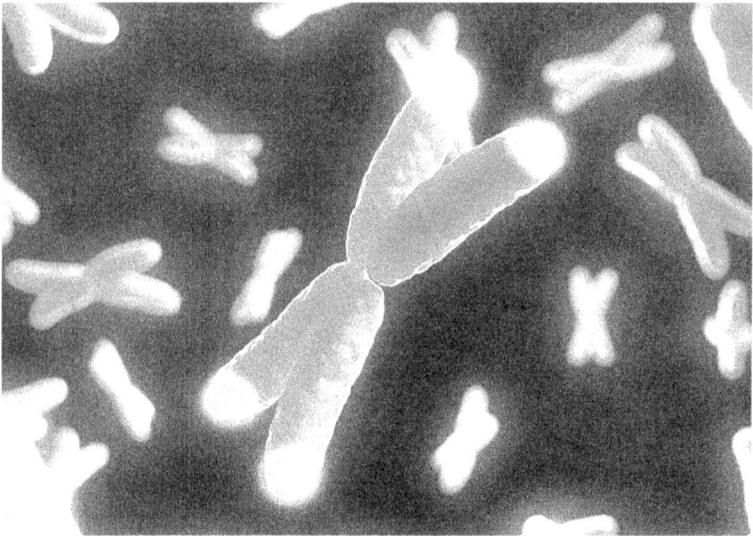

Fig. 12 - Telomore location on the end of a chromosome.

We do not want to get too technical here, but you can visualize the plastic piece on the end of a shoelace. The technical name for the plastic sheath on the end of the shoelace is an aglet. Just as the aglet is important for the functioning of the shoelace, it is the same for the telomere on the end of the chromosome. The telomere keeps the chromosome healthy and if it shrinks, it affects the integrity of the genes on the chromosome. Short telomeres can be equivalent to a poorer quality of life.

Telomere length has emerged as a marker of exposure to oxidative stress and aging. Race/ethnic differences in telomere length have been frequently investigated. In a study by Diez Roux et al. (2009), leukocyte telomere length (LTL) was assessed in 981 white, black and Hispanic men and women aged 45-84 years in the Multi-Ethnic Study of Atherosclerosis. On average blacks and Hispanics had shorter telomeres than whites. Blacks and Hispanics showed greater differences in telomere length associated with age than whites. Differences in age associations were more pronounced and only statistically significant in women. The authors conclude that race/ethnic differences in LTL may reflect the cumulative burden of differential exposure to oxidative stress over a lifetime.

More convincing evidence was found in the Jackson Heart Study by Jordan et al. (2019). They evaluated the association of psychosocial factors (negative affect and stressors) with LTL in a large sample of African American men and women (n = 2,516). Using multivariable linear regression, the authors examined the sex-specific associations of psychosocial factors (cynical distrust, anger in and out, depressive symptoms, negative affect summary scores, global stress, weekly stress, major life events, and stress summary scores) with LTL. Model 1 adjusted for demographics and education. Model 2 adjusted for model 1, smoking, alcohol intake, physical activity, diabetes, hypertension, and high-sensitivity C-reactive protein. Among women, high (vs. low) cynical distrust was associated with shorter mean LTL in model 1 (b = -0.12; p = 0.039). Additionally, high (vs. low) anger out and expressed negative affect summary scores were associated with shorter LTL among women after full adjustment (b = -0.13; p = 0.011; b = -0.12, p = 0.031, respectively). It was also found that high levels of cynical distrust, anger out, and negative affect summary scores may be risk factors for shorter LTL, particularly among African American women.

In men, similar results have been found by Shrock et al. (2018). African American men in the USA experience poorer aging-related health outcomes compared to their White counterparts, partially due to socioeconomic disparities along racial lines. Greater exposure to socioeconomic strains among African American men may adversely impact health and aging at the cellular level, as indexed by shorter LTL.

This study examined associations between socioeconomic factors and LTL among African American men in midlife, a life course stage when heterogeneity in both health and socioeconomic status are particularly pronounced. Using multinomial logistic regression, a sample of 92 African American men between 30 to 50 years of age were studied.

Shrock et al. (2018) concluded that greater financial strain was associated with higher odds of short versus medium LTL (odds ratio (OR)=2.21, p = 0.03). Higher income was associated with lower odds of short versus medium telomeres (OR=0.97, p = 0.04). Exploratory analyses revealed a significant interaction between educational attainment and employment status (χ^2 = 4.07, p = 0.04), with greater education associated with lower odds of short versus long telomeres only among those not employed (OR=0.10, p = 0.040). Concisely, cellular aging associated with multiple dimensions of socioeconomic adversity may contribute to poor aging-related health outcomes among African American men.

Gebreab et al. (2016) have shown that the neighborhood and living conditions can be influential on health outcomes, particularly in women. It is believed that African Americans (AA) experience higher levels of stress related to living in racially segregated and poor neighborhoods, so this study examined whether perceived neighborhood environments were associated with telomere length (TL) in AA after adjustment for individual-level risk factors.

The analysis included 158 women and 75 men AA aged 30-55 years from the Morehouse School of Medicine Study. Relative TL (T/S ratio) was measured from peripheral blood leukocytes using quantitative real-time polymerase chain reaction. Women had significantly longer TL than men (0.59 vs. 0.54, p=0.012). After controlling for sociodemographic, and biomedical and psychosocial factors, a 1-SD increase in perceived neighborhood problems was associated with 7.3% shorter TL in women (Mean Difference [MD]=-0.073 (Standard Error=0.03), p=0.012). Overall unfavorable perception of neighborhood was also associated with 5.9% shorter TL among women (MD=-0.059(0.03), p=0.023). Better perceived social cohesion were associated with 2.4% longer TL, but did not reach statistical significance

110

(MD=0.024(0.02), p=0.218). No association was observed between perceived neighborhood environments and TL in men. The authors suggest that perceived neighborhood environments may be predictive of cellular aging in AA women even after accounting for individual-level risk factors.

Beyond neighborhood investigations, discrimination in general was investigated by Chae et al. (2016). Racial discrimination, a qualitatively unique source of social stress reported by African American men, in tandem with poor mental health, may negatively impact LTL in this population. The study by Chae et al. examined cross-sectional associations between LTL, self-reported racial discrimination, and symptoms of depression and anxiety among 92 African American men 30-50 years of age. LTL was measured in kilobase pairs using quantitative polymerase chain reaction assay. Controlling for sociodemographic factors, greater anxiety symptoms were associated with shorter LTL (b=-0.029, standard error [SE]=0.014; p<0.05). From these findings, the authors highlight the role of social stressors and individual-level psychological factors for physiologic deterioration among African American men. Consistent with research on other populations, greater anxiety may reflect elevated stress associated with shorter LTL. Racial discrimination may represent an additional source of social stress among African American men that has detrimental consequences for cellular aging among those with lower levels of depression.

In this final study to review, we will report on a controversial topic that seems to be escalated in the news on a daily basis (i.e., police brutality). Perceived unfair treatment by police, race and telomere length were explored in a Nashville, TN community-based sample (McFarland et. al., 2018).

According to McFarland et al. (2018), police maltreatment, whether experienced personally or indirectly through one's family or friends, represents a structurally rooted public health problem that disproportionately affects minorities. Researchers, however, know little about the physiological mechanisms connecting Unfair Treatment By Police (UTBP) to poor health. Shortened telomeres due to exposure to this stressor represent one plausible mechanism. Using data from a community sample of black (n = 262) and white (n = 252) men residing in Nashville-Davidson County, we test four hypotheses: (1) Black men will be more likely to report UTBP than white men; (2) those reporting UTBP will have shorter telomeres than those not reporting UTBP; (3) this association will be more pronounced among black men; and (4) these hypotheses will extend to those who report vicarious UTBP. The results revealed support for all hypotheses, and the implications for these findings were discussed as they pertain to debates on policing practices and health disparities research.

Conclusion

It is extremely fascinating to review most of the literature exploring melanin properties and DNA damage coming from melanin-recessive researchers in places like Belgium, France, Germany, Italy and Slovenia. Many scientists from European countries are frantically seeking a solution to prevent aging, and we are sure it is not to assist people of color. It is a full-court press to find ways to ward off the impending doom of the global climate crisis. The planet is heating up and melanin-recessive people are at great risk for negative health consequences.

In the value of the research coming from Europe, they are proving the biological advantages of a melanin-dominant system as a superior biological system to maintain health. This fact cannot be refuted, so we must move toward eliminating self-hate in melanin-dominant people who desire to lighten their skin to be like the melanin-recessive researchers who are trying to find scientific ways to darken their light-colored skin. The world is truly upside down.

In North American research centers, we have presented numerous convincing studies by scientists from all over the USA who have investigated the effects of stress on DNA damage in nonwhite people. Researchers have collaborated from different universities to investigate the link between psychosocial stress and microaggressions on telomeres. The experimental findings are valuable as we explore ways to reduce stress and death for melanin-dominant people. It is not research to simply study skin pigmentation and changes in melanin. The research from these North American scientists provides proof and clear evidence for epigenetic factors affecting the overall health of melanin-dominant people. Melanin may be protective as an antioxidant, but the double-edge sword can attribute to damage in the genes of a person living under constant psychosocial stress. In the end, microaggressions can damage DNA and lead to a person having a poorer quality of life.

CHAPTER 10

TECHNOLOGY AND
THE BLACKEST OF BLACK

Blacker than black is still black.

The search for materials to function like melanin have been explored in numerous carbon-based products. Graphene is one material item that has the potential to revolutionize the scientific application of melanin-related products. Decades ago, Barnes (1988) made a profound conclusion in his book called *Melanin: The Chemical Key to Black Greatness* about the significant implications of studying melanin. Barnes stated that "the melanin molecule with its multifunctional properties show many avenues for the commercial development of products that may be appropriate for medical use as well as products for use in the home, office, automotive industry, chemical industry, capturing solar energy, fashions, research and development; the potential is limitless. To develop the many areas where melanin may have commercial appeal will require in-depth attention by chemists and other researchers." (Barnes, 1988, p. 89). Even without the discovery of graphene, Barnes was one of the first chemists to discuss the battery-like properties of melanin.

Henceforth, a team of researchers from Carnegie Mellon University (Kim et al., 2016) uncovered that the chemical structure of melanin on a macromolecular scale exhibits, amongst other shapes, a four-membered ring—in other words, a chemical structure that may be conducive to creating certain kinds of batteries based on natural melanin pigments. Functionally, different structures of melanin have quite different chemistries, so putting them together is a little like solving a jigsaw puzzle, with each molecule representing a puzzle piece. You could take any number of the pieces and mix and match them or even stack them on top of each other.

According to Kim et al. (2016), there are several possible configurations for melanin with each having a different function depending on its chemical structure. When these molecules bind to form a macromolecular structure, or a polymer, these polymers can be arranged to create a potential battery material. Based on the readings the researchers gained from their experiment, they discovered that a tetramer structure, a four-membered ring composed of larger molecules, appears to be consistent with the structural model of melanin macromolecules.

They were able to discover the tetramer structure of melanin by using it as a battery's cathode. However, in doing so, they also discovered that melanin exhibits a two-voltage plateau characteristic of normal battery materials, while outputting a surprisingly high voltage. To the aforementioned researchers it was surprising, but Barnes (1988) had already presented this decades earlier. We had already known that this material from biology could function potentially as a very good cathode material.

Graphene: The Carbon-based Game Changer in Science

With an investigation into nanoparticle technology, graphene has been a product with many applications. Graphene is not graphite. Graphite is a lighter product that is crystalline carbon, and it occurs as a mineral in some rocks. It occurs naturally, and it is a stable form of carbon. In contrast, graphene is a super metal with limitless uses. Graphene is a crystalline allotrope of carbon with 2-dimensional properties. Its carbon atoms are densely packed in a regular atomic scale chicken wire (hexagonal) pattern.

As reported by reporter Brian Heater, in 2017, a British team of researchers from The University of Manchester marketed running shoes using graphene technology. The research team explored the use of this one-atom-thick material in footwear and then teamed up with a British sportswear brand inov-8 to bring graphene to footwear. Since graphene is the thinnest material available and about 200x stronger than steel, the researchers heated it and added tiny particles to the sole of shoes. When added to the rubber used in inov-8's G-Series shoes, graphene imparts all its properties, including its strength. The unique formulation makes these outsoles 50-percent stronger, 50-percent more stretchy and 50-percent

116

more resistant to wear than the corresponding industry standard rubber without graphene. The Manchester researchers have long discussed graphene's potential role in wearables. In addition to all of the aforementioned "super powers," it's also transparent and more conductive than copper — all great potential traits for the next generation of electronics.

In a review of the literature, graphene research has been useful for the advancement of many scientific enterprises. To make a gamma ray, graphene sheets could be rolled up into a funnel and allow positronium to tunnel through. Since graphene is impervious to all ordinary matter atoms, a laser beam could be created for quantum computer applications. Graphene could be used to reorganize computer chips and help transistors process and store information. It could be used as a heat-shield for cellphones and/or laptops or any temperature sensitive gadget. Studies have also shown the potential of using graphene sheets to block the signals mosquitoes use to identify blood meal and aid in a chemical-free treatment to ward off mosquitoes.

For space travel, vehicles moving at hypersonic speeds are bombarded with ice crystals and dust particles in the surrounding atmosphere, making the surface material vulnerable to damage such as erosion and sputtering with each tiny collision. Using graphene, researchers studied this interaction one molecule at a time to understand the processes, then scaled up the data to make it compatible with simulations that require a larger scale. One application for this work is for research on how to design thermal protection systems for "slender" vehicles and small satellites.

In 1995 and 2004, I discussed the potential marketing of products to exploit melanin. As an example, skin lightening as well as skin darkening agents have been created for world-wide marketing. In addition, the semi-conductive properties of melanin have been used for technological advancements in transportation. With the presence of graphene and the exploration of interstellar metals, this body of knowledge has been advanced.

I once asked the question about the alternative and varying conception to call melanin an electronic-ionic hybrid versus the old view as an amorphous organic semiconductor to a brilliant colleague named Dr. Kelvin Suggs in the Atlanta University Center. When we say hybrid, we are speaking like hybrid vehicles that interchange between gas powered and electric operation to save energy. The properties of melanin can also be used to conserve energy and make biological and synthetic systems function more efficiently.

In a conversation with Dr. Suggs, he stated that the different conception depends on the type and configuration. For example, when a molecule is amorphous then it is typically a large conglomerate or large mass chain with no particular symmetry (i.e., slurry). Bonding is ordered along chains, but the chains tend to be random or often times cross-linked depending on the polymer. The electronic-ionic hybrid is most likely a synthetic form of melanin that is synthesized with higher control, and more symmetry involved such that you get more efficient/ controllable electrical conduction. Suggs further pondered that the electrons like to flow via symmetrical media as those media are typically the path of least resistance (Suggs, Person and Wang, 2011), whereas "amorph"ous scenario has to do with molecules resistant to any one particular "morphology."

Suggs suspects that amorphous systems tend to be of lower conductivity than highly symmetric ones like graphene (Suggs, Reuven and Wang, 2011). Therefore, electronic-hybrid systems may be a maximization of the conductive property. From this conversation, I have sought funding to find biocompatible applications to use graphene.

Several years ago in 2013, I proposed the following project to explore the biomedical applications for the use of graphene and/or nanogold particles in the field of nanotechnology. The use of an available Transmission Electron Microscope (TEM) would have been an excellent tool to uncover the biological properties of nanoscale elements that can be potentially useful for applying graphene or nanogold to the nervous system. It has been reported that graphene suffers from low solubility in aqueous systems which results in their limited biocompatibility. Therefore, current research is needed to explore ways to improve the water solubility of nanoscale elements and to uncover physical and chemical properties that can be manipulated to allow products such as graphene to efficiently interact with biological systems.

Gold nanoparticles (AuNPs), in contrast to graphene, have better properties to affect biological systems. In general, AuNPs can have excellent catalytic activities (Cheng et al., 2012; Msezane et al., 2012). The metallic nature of AuNPs allows these nanoparticles to catalyze oxidation, hydrogenation and hydrochlorination reactions when supported by oxides or carbon substrates (Orive et al., 2011). The use of the TEM could aid in the generation of a new research exploration. Combining the technical expertise of chemistry, biology and other STEM fields can help discover the optimal physical and chemical properties that can make graphene biocompatible (Liu et al., 2012a; 2012b; Zhang et al., 2012a; 2012b; Feng and Liu, 2011).

Graphene has attracted a considerable amount of attention due to the ease in isolating a single sheet of graphite via mechanical exfoliation. Despite the fact that it is one atom thick, exfoliated graphene has shown extraordinary electronic, vibrational and optical properties that can be used as a novel material for many potential applications. When you combine this with gold, the potential is tremendous. Among metallic nanoparticles, gold nanoparticles (AuNPs) have attracted considerable attention in both ancient and present research. Ancient scientists manipulated gold and understood the importance of carbon-based elements. It is suspected that a similar technology was used in the construction of the pyramids, for example, via the manipulation of energy.

The Technology and Blackness

When I visited Kemet, I was amazed at the black granite cap-stones for the pyramids (Figure 13a and 13b).

Fig. 13 - Dismantled black granite benbens in Cairo Museum. The benben was the primeval mound, the first to catch the rays of the sun as caps on pyramids. The cap is also called the pyramidion.

In Kemet, a capstone or pyramidion was called the benben. The benben was the primeval mound, the first to catch the rays of the sun as caps on pyramids (Bunson, 2002). The benben was not black just to be black. In the museum, I could touch and feel the benben, but I could not lift the benben. By whatever means these massive objects were lifted, it was quite obvious that they were placed at the top of the pyramids to conduct energy from the sun. Gold was used in various ways, so the combination may have provided the semiconductive elements to move large objects and transmit energy.

For modern technology, it costs 1.2 billion dollars to make a Stealth Bomber for the military. As scientists have searched for the blackest of black, these materials could be used to block radiation and have serious implications for stealth and defense. The cost is astronomical for one plane. If it costs 1.2 billion for one plane, Stewart (1996) has speculated on how much it would be to equate the cost for the labor, ingenuity, equipment and material to build the Great Pyramid (Stewart, 1996).

Iron had to be one of the elements used to construct ancient monuments. Iron, one of the most abundant elements on Earth, has an origin that is astronomical. By scientific standards, the origin of iron is one of the most violent processes imaginable, and it starts from the explosion of stars. A type of star known as a red giant begins to turn all of its helium into carbon and oxygen atoms. Those atoms begin to turn into iron atoms, the heaviest type of atom a star can produce. When most of a star's atoms become iron atoms, it becomes what is known as a supernova. It explodes, showering space with iron, oxygen and carbon atoms far and wide.

Black people in Africa did not wait for "white saviors" to come tell them how to use nature to build civilizations and construct massive monuments. The question is: "Did interstellar contact help influence these ancestors to use this dark element (i.e., iron) here on Earth?"

Ben Jochannan (1989) provided evidence for the scientific discovery of smelting iron ore as far back as 43,000 B.C. Deep in southern Africa, the evidence for an iron mine in Swaziland was radio-carbon dated to a prehistoric time when melanin-dominant people were in the cavern of mountains using stone-age mining tools to mine hematite – a source of iron. These early African miners had excavated for hematite rich in specularite – one of the prized pigments and cosmetics of ancient times. Furthermore, samples of charcoal were found. If you have ever seen the blackness of coal and heat, the use of it can change the course of humanity. Warmth and heat can make you peaceful.

In contemporary science, researchers claim to have made the darkest material on Earth, a substance so black it absorbs more than 99.9 percent of light (Steenhuysen, 2008). A research team from Rice University in Texas made the finding in 2008. This material, made from tiny tubes of carbon standing on end, is almost 30 times darker than a carbon substance used by the National Institute of Standards and Technology as the current benchmark of blackness. All light is absorbed, and it is pushing the limit on how much light can be absorbed into one material. The substance has a total reflective index of 0.045 percent – which is more than three times darker than the nickel-phosphorous alloy that has the record on the world's darkest material. Basic black paint, by comparison, has a reflective index of 5 to 10 percent.

As reported by Steenhuysen (2008), a team of researchers from Rensselaer Polytechnic Institute in Troy, NY has also worked on this research and they report the material gets its blackness from three factors: (1) Tiny nanotubes of tightly rolled carbon that are 400 times smaller than the diameter of a strand of hair absorbs light; (2) These tubes are standing on end, much like a patch of grass, and this arrangement traps light in the tiny gaps between the blades; and (3) The researchers have made the surface of this carbon nanotube carpet irregular and rough to cut down on reflectivity.

Research on the blackest of black is a constant moving target. The race has been on to achieve blackness because another team from the Massachusetts Institute of Technology (MIT) boasts of even greater blackness (Chu, 2019). In 2019, MIT engineers reported that they have cooked up a material that is 10 times blacker than anything that has previously been reported. The material is made from vertically aligned carbon nanotubes, or CNTs — microscopic filaments of carbon, like a fuzzy forest of tiny trees, that the team grew on a surface of chlorine-etched aluminum foil. The foil captures at least 99.995 percent of any incoming light, making it the blackest material on record.

Blackest of Black and the Celestial Connection

We know coal is black, but could there be the existence of coal black planets? There are numerous types of planets that exist in the dark matter or cosmic cohesion of the universe. In another 2008 finding, an April discovery from the Hubble Space Telescope found a melanin-dominant, super black, charcoal dusted planet in the sky. In the deep depths of space, it would be virtually invisible (Trosper, 2015). However, the carbon-rich planet was detected as an exoplanet, and it was named WASP-12b. It cycles around an intense star called WASP-12, and it is in the Milky Way galaxy. Scientist call this black planet (i.e., WASP-12b) "The Planet of Diamond Death." We may not know what it is made of, but it has been discovered that the planet is the blackest of blacks. Unfortunately, melanin-recessive scientists have already given it a negative connotation by relating it to death. WASP-12b reflects so little sunlight that not even freshly-laid asphalt can compare. Researchers are learning more about the planet, but it is so incredibly dark that it almost completely blends into the darkness of space.

This pitch-black planet is not habitable, but its presence highlights the power of pigment or melanin on an intergalactic level. Disappointedly, our use of the term melanin is very limiting. A new conceptualization of blackness is forthcoming. The same way WASP-12b is responding in outer space is the same way our melanin-dominant body is responding in inner space. We know why humans are black/dark in pigmentation, but why would a planet be pitch-black? Researchers have theorized that the planet's close proximity to its star (i.e., WASP-12) could be responsible. WASP-12b is twice the size of Jupiter and it is close to its sun-like star. It completes a full rotation in about one Earth day. That short distance suggests that the planet's surface could top 2600 degrees Celsius. In effect, the atmosphere is responsible for absorbing so much of the intense light that hits the planet. Now that is pigment power.

Astonishingly, there was an out of the ordinary cosmic event that occurred during the late Fall of 2017. It was the fascinating passing of an object zooming at 98,400 mph through our solar system. If it hit Earth, it could have been a major event that could have shut down the global economy and decimated populations on this planet. It could have been more devastating than the 2020 Coronavirus Pandemic (see Chapter 12) which will be discussed. Do you think we would go back to watching a television show after such a cataclysmic event? Our whole world could be upside down in a matter of seconds. Instead, many people do not know that a 400-meter-long (i.e., slender) and 40-meter-wide space object was visiting our solar system. Slender is highlighted again because of what was previously mentioned about designing thermal protection systems for slender vehicles with graphene.

To be specific, this interstellar space rock was not circular in any manner, and it had a hyperbolic trajectory path. On September 9, 2017, it was close to the sun; October 14 it was close to the Earth. From an observatory in Hawaii (University of Hawaii's Pan-STARRS1 telescope), it was first spotted October 19, 2017 before it eventually zoomed away from our solar system.

In its hyperbolic path, it came extremely near our sun and it did not burn up. It had a physical/chemical composition that did not allow it to be burnt and sizzled by the sun's solar radiation. The oblong structure was from another galaxy far away. It was so different, astrophysicists had to give it a new classification. It was neither a comet [C] nor an asteroid [A], so it was given [I] for interstellar. The technical name was 1I/2017 U1. It was named Oumuamua which is Hawaiian for "messenger from afar arriving first." Was it a spaceship? We might never know.

Cosmic Seed of Life

If there was any believable discussion of spaceships, other intelligent life in the universe, and other planetary systems, then the African people in Mali would be where the evidence could be found. How, without telescopes, could the Dogon know that 51 trillion miles away (8.7 light-years), the orbit of an invisible star around its companion was 50 years duration? Could their darkly pigmented eyes and melanin-dominant bodies assist with their direct connection to the cosmos? Has being isolated and away from the contaminated thoughts of Eurocentric thinking potentially elevated their consciousness to contain what was destroyed from past civilizations? We do not know exactly, but I can bet you there were not little green men with big eyes that came down. With the monopoly on knowledge, we need to question the many "alien" abductions that some white people have allegedly experienced with green space people. Perhaps this has been a scheme to distract melanin-dominant people who really are connected to the universe.

The extensive details are written about the Dogon (Griaule and Dieterlen, 1986; Adams 1983a; 1983b; Finch, 1998), but the connection we are highlighting here pertains to melanin-recessive people at the frontier of space exploration. Finch (1998) titled his fabulously written and extensively referenced book, *The Star of Deep Beginnings*, to reflect the seed star (po tolo). It is not visible to our weak naked eye, but we cannot say it is black and virtually invisible like WASP-12b. All the speculative connections of this star made by the Dogon pertain to it as the place all life on Earth came from. It is pigment power on cosmic display.

The star known as Sirius B is found in the Orion star system. The Dogon call Sirius B, po tolo, and its companion Sirius is called sigi tolo. It is assumed the Dogon have known this in their tradition for hundreds of years and they know Sirius B is incredibly dense and heavy. In fact, modern science has shown it is the densest and heaviest of the non-neutron stars. It is the mother seed of all stars and one of the oldest heavenly bodies (10 billion years or more). The universe is 13.8 billion years old, and Sirius B was originally sun-sized until it fizzled towards the end of its life span and turned into a white dwarf. The substance of a white dwarf gradually converts from a diamond-like carbon material to a metal, like iron but different from it. Both Sirius B and WASP-12b are filled with carbon, an element for life.

The death of a white dwarf can make possible the birth of other stars and star systems. In fact, all newborn stars are created when older stars die, so our solar system was created out of the dust of a dying star such as Sirius B. The Dogon believe Sirius B was once closer to us versus the current position, and they declare that our solar system – including Earth – owes its very existence to Sirius B (Finch, 1998).

In the next chapter, we will embark on an astrobiological journey to suggest that the saying "ashes to ashes and dust to dust" has some ancient relevance to human beings as a representation of the full manifestation of star dust. We know we have all the elements of Earth in our bodies, and the melanin (a carbon-based molecule) can be the trigger agent to keep us connected to the creative force of the cosmos. Pigment power is the conduit between the material-spiritual connection (Moore, 1995; 2002; 2004). Given this conjecture, we do need to investigate the real possibility of travelling through space and time, mentally and physically.

CHAPTER 11

ASTROBIOLOGY AND MELANIN

It is the mind that holds the destiny for every entity in the cosmos.

The major theoretical viewpoints in western culture pertaining to the origin of the universe have been The Creation Theory, The Theory of Evolution and Intelligent Design. Beyond these major themes, we must give acknowledgement that many cultures throughout the world have their own conceptualization of how the universe was formed and how man/woman came into existence. In common with virtually all non-Western peoples, Africa has, from the beginning, conceived of the universe as a mental expression of the creator (Finch, 1998). We do not want to claim to have the final answer in this book, but the content will be made to analyze our existence from an astrobiological position. Astrobiology is defined as an interdisciplinary scientific field concerned with the origins, early evolution, distribution, and future of life in the universe. It may also consider whether extraterrestrial life exists, and if it does, how can humans detect it.

In the expansive universe, ashes to ashes and dust to dust are how stars are born. Some of the air we breathe may have been released from major cosmic events that make white dwarfs, red giants and supernovas. The dispersal of carbon, silicon atoms and other elements are violently dispersed in the cosmos. Astrophysicists suggest the very materials from which our bodies are made were derived from this violent process somewhere in the universe.

It is interesting that all the major theorizers of the major theories are melanin-recessive people. In other words, western, Eurocentric thinkers have had a major influence on thoughts about the origins of the universe and where humans came from. This white-oriented view is problematic because the lens of the observer will put white-oriented conceptions into the discussion and this colonizer-type of delivery will lead to only white images that matter and all else is irrelevant. This is extremely problematic because melanin-recessive people are a minority on the planet (Welsing, 1991), and there is no way they should be represented as the majority authority on science principles.

For the western, Eurocentric view to have preeminence, it creates a monopoly on knowledge. A prime example of this false indoctrination is how many people from various cultures throughout the world believe that Jesus is a pale-skinned, straight-haired, blue-eyed, melanin-recessive white man. When people are questioned about this belief and the comment is made about Jesus being another color, many religious zealots will say, "well color does not matter." To the contrary, it does matter, and it has everything to do with the control of the mind. There are people who control knowledge, and therefore, they can control the mind of the masses of people.

From Beyond the Stars?

I have a book purchased in 1994 that is called *Blacks: The Race from Beyond the Stars* by Paul D. Duncan (1994). There is an ISBN and a publisher for the book. Why can't you do a simple internet search to find a copy of this book or information about this book? In fact, odd things pop up on an internet search that are irrelevant to the search. I don't know the author and I know nothing about the author on a personal level. However, I found the information in his book provocative, and it should be open for the public to view. In the global world that has been enhanced by the internet and the information age, there should be nothing that a person cannot find, especially if it was only published 30 years ago. Since the book cannot be located from a simple search using the internet, this extraction of information has led me to believe that it has been deliberately removed from view. The current author has written books that he controls for distribution. Even in death, there will be access to *The Science of Melanin or Dark Matters Dark Secrets* on the internet. Unless there is a secret agency that controls the internet and deliberately removes information from the public, all information should be accessible.

The premise made by Duncan (1994) is that esoteric knowledge about the origin of the Black "race" has been hidden and unrecognized for thousands of years. He presents evidence that the Theory of Evolution is a conspiracy. Duncan informs the reader that his book is the

128

"first book by a Black author in this field to reject the basic premise of evolution itself. The theory of evolution is just as unscientific regarding animal life, as it is human. Therefore, rejection of the entire theory becomes necessary." Beyond Duncan, other contemporary scholars have invested their careers and research time exposing the flaws in the theory of evolution (Yahya, 2007).

In Duncan's 143 paged book, there are four chapters that support scientific, historical, archaeological, astrological and anthropological evidence pointing to mankind coming from another planet, possibly even another solar system. It is believed that the first colonist to arrive on Earth used the moon as a giant, hollowed-out spaceship world, to bring people, animals, planets, equipment, technology, etc. to Earth. Since melanin-dominant and melanin-recessive scientists agree that people of a darker hue were the first Homo sapiens to populate the planet, all subsequent ethnic groups and cultures based the ideas for their cultures on the framework setup and designed by people of a darker hue. The esoteric part of this discourse is related to the archaeological, astrological, mythological and religious information that demonstrates a connection between the Moon and Egyptians, Sumerians, Babylonian, Indian, Chinese and other ancient civilizations based on Black culture.

There are many written and oral traditions of ancient peoples, passed on from generation to generation, that human and animal life were brought to the Earth fully formed from some place outside the terrestrial sphere of the Earth. Many people from various ancient cultures firmly believe that the "someplace else" was the Moon. It is interesting to note that some of the first hominid fossils linked to man were found in East Africa at a place called Mount Kilimanjaro (i.e., foothills of the mountain of the moon). If we add Mars to the story, Cairo in Egypt actually means Mars. An even deeper analysis of melanin-dominant people connected to the cosmos is definitively explained by Adams (1983) and Finch (1998). To even conceptualize how a melanin-dominant people deep in the interior of Africa could know about the elliptical patterns of star systems (e.g., Sirius) without telescopes is super perplexing.

Hunter Adams (1983a; 1983b), an expert on the topic, has studied the Dogon for decades. To expound on the importance of space for ancient melanin-dominant scientists, he states that, "all over the earth stand their mysterious, megalithic monuments tuned into the rhythm of the stars, cycles they realized were the pulse of life. Through exacting long-term observation, our ancestors determined that seasonal cycles, vegetation growth cycles, and even animal migrations and mating cycles correlated with the cyclic changing position of the moon and sun." In a phenomenal text by Finch (1998), *The Star of the Deep Beginnings*, he provides definitive evidence for the paleolithic beginnings of formal mathematics, astronomy, engineering, architecture, navigation and map-making by melanin-dominant people. In addition, Finch delves into the interstellar space knowledge of the Dogon.

The Dogon People of West Mali have profound information about an existence coming down from above. In *The Pale Fox* by Griaule and Dieterlen (1986), it states, "that a basket intended for carrying things represent, when turned upside down, the ark on which humanity descended from heaven to earth, the square bottom of the object representing space and the cardinal points." p. 64. Interestingly, Enoch, from the Book of Enoch, was criticized for viewing the world as squared flat with four corners. We will discuss Enoch later, but the point is, ancient people conceptualized the world without the telescopes and microscopes we have today. Is it possible there was information about the cosmos that was provided from above? Is the basket or ark reflecting on a space vehicle? The Dogon further make the comment below about creation, and it sounds like they are describing DNA before the Nobel Prize was won by Watson and Crick in 1962 for revealing the double helix spiraling structure of DNA.

The Dogon believed man is a combination of seeds, symbols of the life forces. According to Dogon philosophy, "the present world is conceived as having come out of a first seed formed by God, this being *Digitaria exilis*, the fonio. It contains the essence of creation, including the four basic "elements" (air, earth, water and fire) and the "word" of the creator, that is to say, life manifesting itself within, in the form of eight segments, animated by a motion that is both vibratory and spiraling." p. 63. Now for the reader, this is describing creation and intelligent design and science.

The word is the thought coming from the mind that creates all. The ALL is the first law of the Kybalion by the Three Initiates (1988). From a biblical perspective, in the beginning was the "word." To expound here, the word of creation makes the seed we know of as DNA that has four basic pairs of nucleic acids: Adenine, Thymine, Cytosine and Guanine. The eight segments mentioned by the Dogon are the DNA structure of these nucleic acid pairs. The outer supports of the DNA ladder are made up of molecules of deoxyribose sugar alternating with bonds of phosphate. The connections between these supports are made up of two types of paired nucleic acid bases that vibrate and spiral. From this author's perspective, they were describing genetics in their cosmology.

Duncan (1994) makes the argument that if man "evolved" from apes, why can't a human mate with them and produce a half-man, half-ape offspring? He highlights major differences between man and ape, and a summary of his major points are:

1) Man has a flexible hand and sensitive fingertips;
2) Humans have soft skin and a low healing rate;
3) Man enjoys food and swallowing slowly;
4) The helplessness of the human infant and long childhood of humans; and
5) Man's inability to mate with his "closest relative."

On this last point, it is factual that man is unable to mate and produce offspring with the one mammal (the anthropoid apes) that scientist claim is man's closet relative. Even though we know Homo sapiens share 99% of DNA with Chimpanzees and Bonobos, there is no hybridization between human stocks and anthropoids of prehominid types. Interestingly, humans share 99.5% of DNA with Neanderthal, and 20% of Neanderthal DNA is in White and Asian people, not Black/African people (Perritano, 2019).

In a bizarre reflection on DNA, mice and humans share 85% of their DNA and humans have fewer genes than a nematode worm. However, in the creation and formation of man, it should be noted that only 1% of genes in DNA carry instructions to make proteins. The rest (i.e., 99%) is so-called "junk" DNA. This is a pretty significant fact considering the complexity of the human body and our brains.

As a neuroscientist, I support the view that the brain is the most elaborately constructed entity in the universe. It can create the cosmos and the cosmos created it. With all the junk DNA we possess, it is there for a reason. It is like the bulk of the universe (Dark Matter and Dark Energy) that we cannot see. The stars, galaxies and nebulas are only 0.4% of what we can see in the night sky. Esoterically, humans are a microcosm of the macrocosm, and virtually, miniature replicas of the universe. Since humans are seeds from the stars, the key to unlocking the universe is to know thyself.

The Brain is the Universe

Space is about speculation. The universe is approximately 13.8 billion years old. The Earth is 4.5 billion years old and the moon is 50 million less than the Earth. Radiation has been present for billions of years, so we know melanin is ancient. The carbon-based molecule known as melanin is found in interstellar space and throughout the human body. Melanin can be found in the stars, on the surface of the human body, in the internal organs and deep inside the brain. Although the melanin is formed differently in all of these places, the common link is that all types of melanin are carbon-based and can be intricately connected on a cosmic level. Black absorbs light and energy and this allows for melanin to serve as a fantastic conduit. On a cosmic level, the human body is a microcosm of the macrocosm, and the brain is the key to the link (Figure 14).

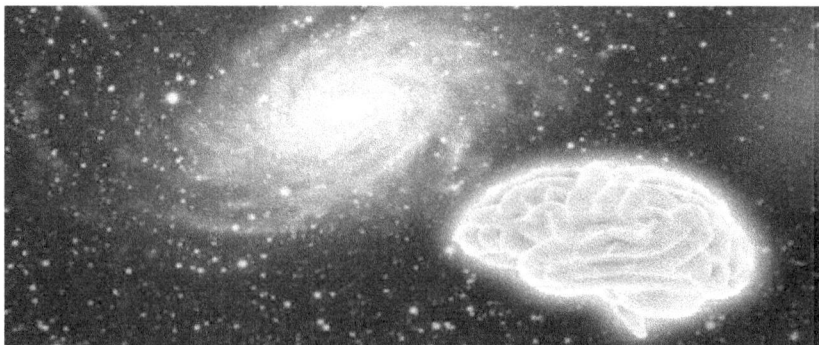

Fig. 14 - Cosmic connection between the human brain and the universe.

132

With the evidence presented, we can suggest that the original people on this planet classified as Homo sapiens who did not evolve from apes, probably have more in common with cosmic beings than apes. A compelling truth is found in the functioning of our brain. All of the animals on earth had the same amount of time to develop intelligence like man, but none did. Man's brain is too big to support a theory of evolution. Most animals, like apes, eat, survive and procreate as an everyday existence. The brain of man, however, has scientific genius, engineering skills, artistic ability, musical talent and mathematical capabilities to build and create.

If we view classical evolution, it makes no sense. For example, all changes that result in a new species requires millions of years. Apes developed from lower hominids to develop 1 billion neurons that make up the anthropoid brain. This is remarkable if you believe in evolution that the earliest life forms (during the Cambrian period) took 500 million years to develop the ape brain. According to Duncan (1994), since man is assumed by evolutionists to have developed from a branch of the ape family, there is a math problem. Man's cortex has 12-15 billion neurons, which meant it should have taken ten times as long to develop as the ape with his 1 billion. Ten times 500 million is 5 billion years.

This is impossible because the age of the Earth is 4.5 to 5 billion years. Based on evolutionary dating, man has not had time to develop his incredible brain. Of all the living creatures on Earth, only man has the ability to improve his knowledge. Only man has the ability to build on accumulated knowledge and there is virtually no limit to man's learning potential. Along with the 99% junk DNA that is not used in humans, it is a common scientific belief that man only uses a tenth of his brain's ability. One ability of the brain that has been both a blessing and a curse is the development of religion.

Religion has been used as an opiate for the masses. Not that it started with ill intentions, but we must be critical since people kill in the name of religion. God or the creative force cannot be the author of such confusion. Is there a reason why man needs to be connected to something greater than himself? Do animals have religious services? Do animals have this need to communicate with a higher creative force like they are connected to a "divine" plan? The answer is no to these questions because where animals stop, man continues to grow.

Duncan (1994) suggests that man has a genetic imprint that has always allowed him to look to the skies for his "father in heaven." Memory genes passed down (Bynum, 1999) thru thousands of generations (along with the many "god" myths and legends) still retain the recollections of the time space voyaging ancestors that potential brought humans to Earth. If we comprehend this cosmic connection, man has been seeking this connection ever since the ancient ancestors departed this realm. Religion is the manifestation of the program in man to search for the meaning of life, to explore, expand, create, learn and reach a cosmic consciousness through meditation (Bynum, 2012). To be one with the creator is the mission of divine universal consciousness.

In his thorough treatise on the philosophy of Africa during the Pharaonic Period, Obenga (2004) provides linguistic evidence that there is no opposition between "matter" and "spirit" in ancient Egyptian philosophy. According to Obenga, "nature is a whole, within which matter and consciousness are merged…In Egyptian thinking, the objective object is inseparable from the subjective subject…The universe as a whole tends toward organization, and it embraces the totality of all that is, spirit as well as matter." p. 57.

Combining material and spiritual concepts may be difficult for scientists today, but it remains a personal quest to interpret the meaning of this coalesced cohesion of consciousness. In fact, Asante (1999) believes that the ancient Egyptians sought to answer five questions that were important to the pursuit of knowledge:

1) How to describe the indescribable?;
2) How to show the unshowable?;
3) How to express the inexpressible;
4) How to seize the ungraspable?; and
5) How to measure the unmeasurable?

For Asante, the answer to these questions was the very basis of knowing, and for this author, it is a reason this book was written.

When studying the brain, these questions about the pursuit of knowledge can relate to brain functioning. Neuroscientists are still investigating how it works, but the brain can perform higher order functioning and create anything imaginable. Since the brain and universe are connected by the mind, it is the mental universe that is the key to maintaining a harmonious existence. It is believed that multi-verses or parallel universes exist, and this paradigm presupposes innumerable universes in the primal seed of the Big Bang explosion. Because an infinitude of possible universes exists, it is possible for there to exist at least one universe that contains all the right elements, circumstances, and processes necessary to evolve higher structures, including life (Finch, 1998).

The Big Bang and a Parallel Universe Possibility

Where does this divine universal consciousness exist? If we describe how scientist use the Big Bang Theory for the creation of our universe, we can get a bigger picture. It was a moment in time and the common moment was thought to be 13.8 billion years ago (Figure 15).

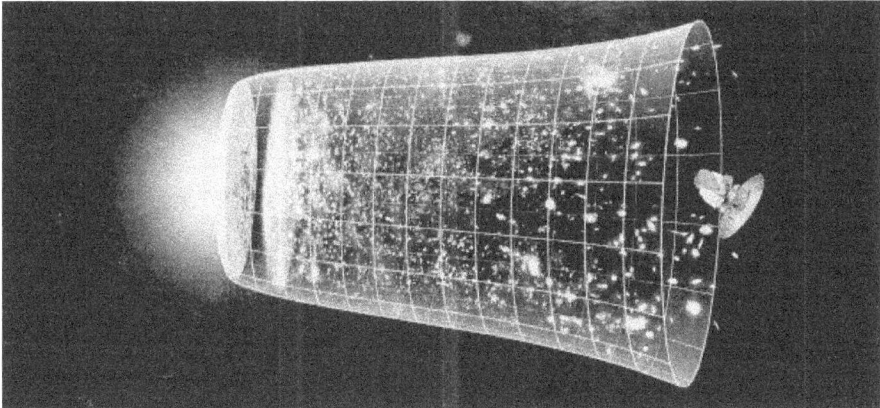

Fig. 15 - The Big Bang and a standard model of the universe with the telescope on a journey at the right.

We can speculate that there was an enormous concentration of energy that exploded, and this singularity in time has been dated back as far as 15 billion years ago by Hawking (2001). Whichever date you choose, it began with a hot and dense time period that was filled with energy. The Dogon even described how stars are moving away from a former position (Finch, 1998), and this can describe the Big Bang concept

135

before it was revealed by modern telescopes. To parallel modern western views, the Dogon have their own version of the singularity of the Big Bang, and they believe in an existence before existence.

What was in existence before the Big Bang? Such an inquiry would make it easy to equate a God-like presence to this beginning, but not from a biblical reference point. We can suggest a creator or creative force was involved and continue with a belief on how the universe began. We can say the Big Bang was not the beginning of time, but a symmetry of a moment in time. The multiple universe theory is a unique speculative point to discuss symmetry and astrobiology.

This author believes that in the intensity of the Big Bang, particles with matching replicas were sent symmetrically in different directions. Therefore, similar carbon-based material was spewed from the explosion in different directions. Since it all came from one source, the infinite cosmos was intimately connected. The symmetry explains how different galaxies formed with different shapes and sizes with matching galaxies (i.e., replicas) out in space.

The divine universal consciousness relates to a mental connection to this cosmic force, and the first principle of the *Kybalion*, states that "The All is Mind; The Universe is Mental." Essentially, the principle embodies the truth that the All or God is Mind. It should have no gender and the principle explains the true nature of energy, power, and matter, along with how these are subordinate to the mastery of the mind (Chandler, 1999).

With the vast expanse of the universe as we know it, there could be a parallel place that is similar to our planet somewhere out there in space. The conditions could be similar or very different, but we cannot close our minds to think this creative force coming from the Big Bang waited 5 billion years later for Earth to form and then 700 million years later on Earth for us to "evolve" from a single celled organism. In this great billion-year span of time, we can speculate that something else is out there in the universe. As we pontificate on other life forms and civilizations, some may have already come and gone.

According the Theory of Evolution, life began in the sea about 3.5 billion years ago from a single-celled organism. This organism developed into more complex, multicellular life forms over a millennium of years. Out there in the cosmos is another form of life, but it might be very different than how we see life on Earth. Given this conception, there is the understanding that we are not alone. There might be multiple histories to parallel with the multiple universes and going through the wormhole in space could send us to another space and time with a different history. In fact, you might meet yourself out there in another place and time in the cosmos. I am sure you have felt déjà vu experiences, so to say we have been here before or seen other experiences before they have occurred may help validate a quantum theoretical view of the universe. There can be more than one future and more than one ending. According to Wolf (1988), it is a situation in which, potentially, a sensitive mind could "remember" the future and "predict" the past. Since past and future co-mingle in the present, quantum physics implies that devices must exist that enable us to tune in on the future and resonate the past. I believe the device is the brain, and as you read now, think about the split-second co-mingling of thoughts in your mind.

The intelligent life need not be anything like humans. In our minds, little black aliens would mentally suffice, and their melanin-dominant bodies would do justice to survive the cosmic radiation that exists everywhere in the universe. Melanin would be an astrobiological material to ensure the existence of life. Even on Earth we can see the pigment power of melanin forming in abandoned nuclear power plants that have leaked. Radiotrophic fungi, for example, were discovered in 1991 growing inside and around the Chernobyl Nuclear Power Plant containing three different melanin-containing fungi. The black fungi grew faster in an environment in which the radiation was 500 times higher than in the normal environment. The fancy names for these black fungi were *Cladosporium sphaerospermum*, *Wangiella dermatitidis* and *Crytococcus neoformans*.

Possessing melanin in radiation charged environments makes biological sense. We can only speculate on what life looks like in another universe, but we are certainly relating a melanin-dominant life form that can adapt to space environments that are highly radiated.

137

Mentioning space travel would seem to suggest there are UFOs that exist. Although we have evidence for UFOs, Hawking (2001) would discount the conspiracy theory that UFOs are from the future and the government knows and is covering it up. For Hawking, he believed the record of cover ups is not good.

Hawking was a brilliant scientist. However, he would have had no problem attributing a white-oriented perspective on interstellar contact. In fact, the use of terms such as black hole, dark matter and dark energy all signify an unconscious Eurocentric thought pattern on black as bad and dangerous. It makes sense for this thinking from melanin-recessive people, but it does not mean it should be the universal language of the masses of melanin-dominant people on the planet. From an African/Black, melanin-dominant perspective, a black hole would be an energy vortex, dark matter would be cosmic cohesion and dark energy would be cosmic breath. Although cake is cake, we must get away from the Devil's Food cake (black) and Angel Food cake (white) thought pattern. When we change our consciousness, we can find different answers. With the massive structures that have been unexplained all over the Earth, the perspective that a higher consciousness existed to impact our presence on this planet is a thought many cannot contemplate. When they do contemplate it, their work (Duncan, 1994) is hidden from the public.

DNA and Planetary Pigmentation

In the Dogon system, we find the Principle of Correspondence manifested throughout their cosmology. The seed stars are equivalent to seed plants. Depending on how you fixate your perception, the universe is man and man is the universe. Moreover, the recognition of the internal system of stars in our solar system relate to the system of internal organs in the human body. With all these layers of correspondence, melanin is present everywhere to spring forward life.

Pigment must be present in all biological systems because it is needed to both process light for photosynthesis and to protect cells from too much ultraviolet radiation. In plants, the pigment chlorophyll is critical and in humans we have continually documented the importance of melanin for life.

Whenever we officially determine the time period when Homo sapiens were first formed/created on Earth, we know they were darkly pigmented to survive in the harsh environment. Humans do not have a thick exoskeleton and humans are not furry. The skin is delicate and exposed, so pigment becomes the critical factor to guarantee survival from the intensity of the sun. Without the power of this natural pigment, the DNA in cells could be altered or damaged and life for humanity would cease to exist.

Defining what a human is and what made him/her has ignited some of the most volcanic and acrimonious debates in the academy. The salient nature of paleontology is controversial because it is a discipline that has much to infer from very little. Therefore, speculation on the cosmos can be just as speculative on where we came from (Figure 16).

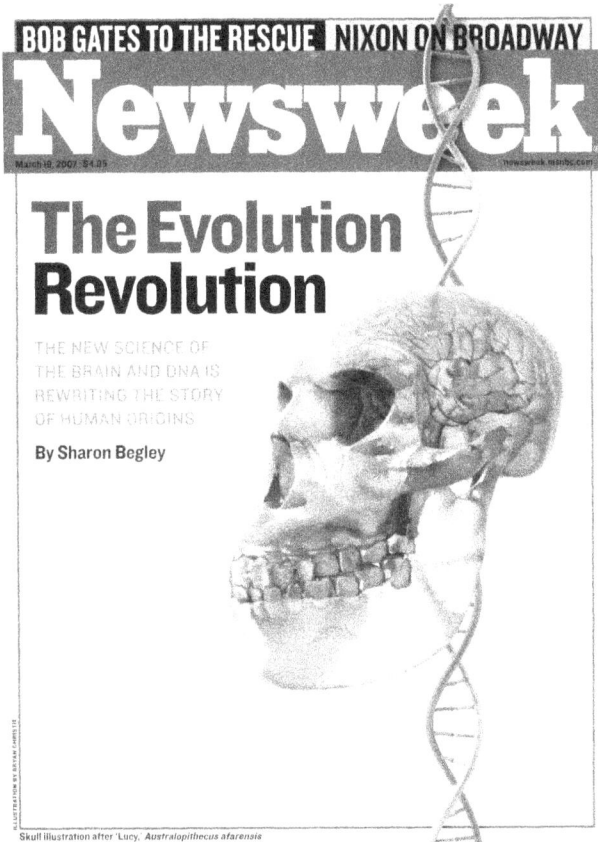

Fig. 16 - Newsweek magazine issue focusing on the constantly changing view of human origins.

If modern man/woman evolved from apes, we would still be mating with the lower species since apes are not extinct. The only reason they may go extinct today is because of man-made destruction to their natural environments as well as poaching. For clarity about appearance, African/Black, melanin-dominant people do not physically display a flat-buttocks, straight hair on the body, thin lips and white skin under the hair of the body. Therefore, the original melanin-dominant people on the planet DID NOT come from apes and monkeys. The only reason racists relate black people to monkeys is due to the dark fur and the supposedly low intelligence of nonverbal monkeys. However, this has been a false, spooky thought in the racist mind that creeps in with doubt about where white people really came from. This book has not been about analyzing the thoughts of white people, but consideration must be given because Eurocentric thought has tainted our conceptualization of the world (Ani, 1994).

To be astropomorphic is to think that other life forms in the cosmos think like humans on planet Earth. It is like anthropomorphic in which humans give human-like qualities to animals. It is a major distraction in the psyche to be concerned with the xenophobic mentality of melanin-recessive, war-like people (Bradley, 1978) who foment war to stay in power. As stated by Orwell in 1984, "War is Peace." More than likely, this type of mentality has nothing to do with the thinking of cosmic beings. Upon their return to our solar system, let us hope they do not make contact with melanin-recessive people first. The consequences will be dire, and history is replete with the horrendous outcomes of melanin-recessive people meeting the unfamiliar. Something is quite backwards in this type of thinking and developing a "Space Force" as the Fifth Branch of the military (Burns, 2019) demonstrates a continuing danger that will hinder world peace. It is highly likely that a kill or be killed mentality will be on display. On Earth, we sure cannot call it intelligence; it is death thinking. It is very doubtful that cosmic beings think this way.

We turn to scientist and history to address the topic of planetary pigmentation. There is no scholar more thorough and worthy of recognition on this topic other than Dr. Cheikh Anta Diop (1974; 1991). Along with his trusted disciple and a distinguish student by the name of Dr. Charles Finch, they have been at the forefront of breaking down the origins of humanity. To support the fossil evidence, I am in full support of the record they present (Diop, 1974; Finch, 1991). It is worthy of reading their intellectual contributions because the scientific evidence presented is overwhelming in support of a rational transition from melanin-dominated Homo sapiens to lighter colored human species. In fact, this view was presented (Moore, 1995, 2004), so there is no need to repeat here. What is different is the addition of evidence presented in this book to discuss an extraterrestrial influence on modern man. The cosmic influence does not take away the fossil progressions and the science behind melanin on planet Earth. Instead, the cosmic influence adds to the story of speculation because we do have an unexplained understanding of the grand scheme of technology that was utilized by early civilizations dating back thousands of years.

(see Table 1 on following page)

Both Diop and Finch provide the best interpretations of the origin of man on Earth and how the Black human populated the planet. There is always conjecture in science, but a foundation on black as the beginning makes the most sense. The common theme from Eurocentric scholars is to belittle the presence of Homo sapiens in Africa. To highlight this point about deception, recall that for close to 40 years, there was a hoax (i.e., Piltdown Man) to think modern man began in Europe. Therefore, we must be careful with the interpretation by white historians. Reading all the works by scholars like Diop and Finch will alter your reality on origins. First, let us quickly summarize the theory of evolution.

Table 1 – A Brief History of Modern Man's Existence

1.75 mya to 12,000 ya [Ice Age to Warming Period]

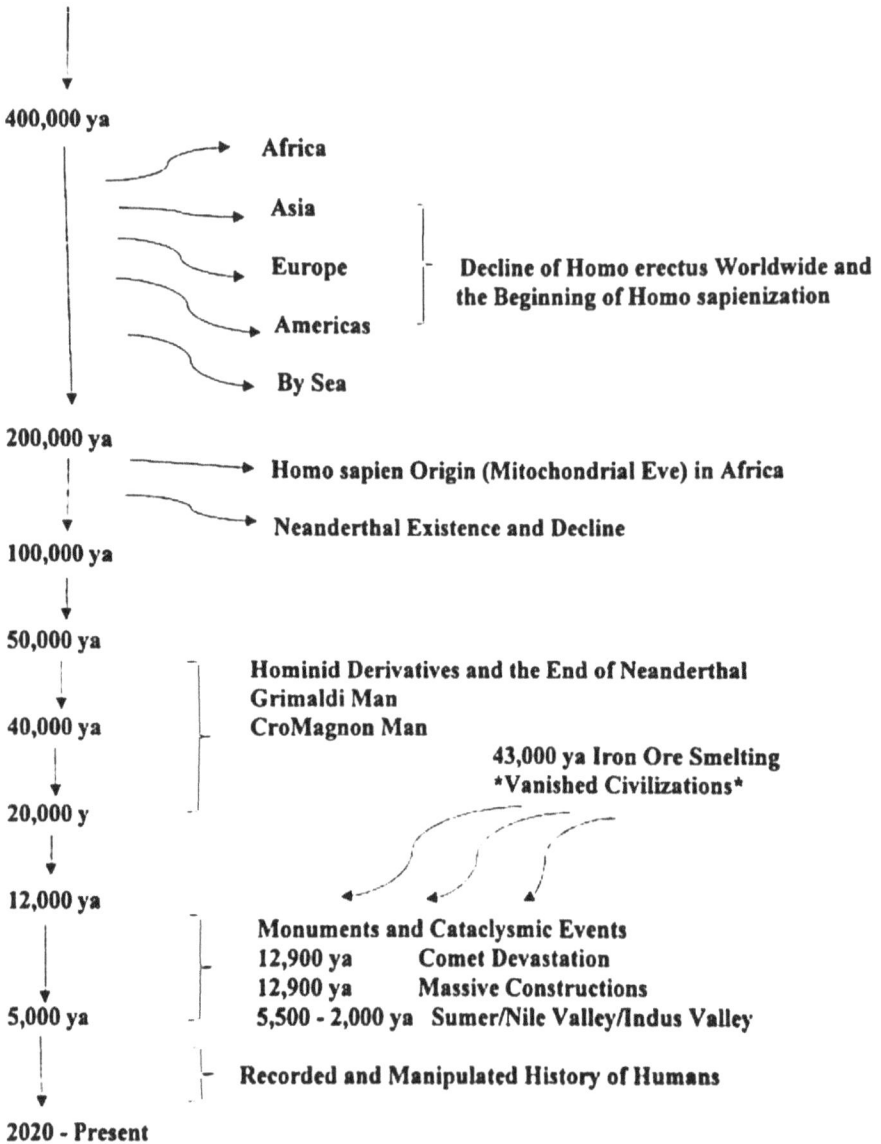

400,000 ya

Africa

Asia

Europe — Decline of Homo erectus Worldwide and the Beginning of Homo sapienization

Americas

By Sea

200,000 ya

Homo sapien Origin (Mitochondrial Eve) in Africa

Neanderthal Existence and Decline

100,000 ya

50,000 ya

Hominid Derivatives and the End of Neanderthal
Grimaldi Man

40,000 ya

CroMagnon Man

43,000 ya Iron Ore Smelting
Vanished Civilizations

20,000 y

12,000 ya

Monuments and Cataclysmic Events
12,900 ya   Comet Devastation
12,900 ya   Massive Constructions

5,000 ya

5,500 - 2,000 ya   Sumer/Nile Valley/Indus Valley

Recorded and Manipulated History of Humans

2020 - Present

Table 1 - Brief History of Modern Man's Existence on Planet Earth.

142

The Theory of the Evolution of Us and the Absence of Melanin

Molecular evidence suggests that between 8 and 4 million years ago, first the gorillas (genus Gorilla), and then the chimpanzees (genus Pan) split off from the line leading to the humans. For a comparison, human DNA is approximately 98.4% identical to that of chimpanzees when comparing single nucleotide polymorphisms. The evolutionary history of the primates can be traced back 65 million years, and one of the oldest known primate-like mammal species, the Plesiadapis, came from North America; another, Archicebus, came from China. Other similar basal primates were widespread in Eurasia and Africa during the tropical conditions of the Paleocene and Eocene. In other words, the primates we are known to evolve from were all over the planet and not just Africa.

Apes existed in many regions of the planet and there are epic periods in history that can be a challenge to follow because new fossils are found every few years. During the Miocene, from 23 to 5 million years ago, a bewildering assortment of apes and hominids inhabited the jungles of Africa and Eurasia (apes are distinguished from monkeys mostly by their lack of tails and stronger arms and shoulders, and hominids are distinguished from apes mostly by their upright postures and bigger brains). The most important non-hominid African ape was Pliopithecus, which may have been ancestral to modern gibbons; an even earlier primate, Propliopithecus, seems to have been ancestral to Pliopithecus. As their non-hominid status implies, Pliopithecus and related apes (such as Proconsul) weren't directly ancestral to humans; for example, none of these primates walked on two feet.

Ape (but not hominid) evolution really hit its stride during the later Miocene, with the tree-dwelling Dryopithecus, the enormous Gigantopithecus (which was about twice the size of a modern gorilla), and the nimble Sivapithecus, which is now considered to be the same genus as Ramapithecus (it turns out that smaller Ramapithecus fossils were probably Sivapithecus females). Sivapithecus is especially important because this was one of the first apes to venture down from the trees and out onto the African grasslands, a crucial evolutionary transition that may have been spurred by climate change.

143

Toumai is the nickname of an old fossil skull, virtually complete primate, discovered by Ahounta Djimdoumalbaye July 19, 2001, in the desert in northern Chad Djurab site TM266. This new hominid is the oldest known (7 million years) representative of the human lineage. It led to the definition of a new species, Sahelanthropus tchadensis, probably very close to the chimpanzee divergence hominins. The hominines are the hominid line linked to the famous fossil line known as Australopithecus (i.e., Lucy) 3.5 million years ago.

Toumai, however, has complicated the story with its more than 7 million years ago presence. Paleontologists disagree about the details, but the first true hominid appears to have been Ardipithecus, which walked (if only clumsily and occasionally) on two feet but only had a chimp-sized brain; even more tantalizingly, there doesn't seem to have been much sexual differentiation between Ardipithecus males and females, which makes this genus unnervingly similar to humans. A few million years after Ardipithecus came the first indisputable hominids: Australopithecus (represented by the famous fossil" Lucy"), which was only about four or five feet tall but walked on two legs and had an unusually large brain, and Paranthropus, which was once considered to be a species of Australopithecus but has since earned its own genus thanks to its unusually large, muscular head and correspondingly larger brain.

Both Australopithecus and Paranthropus lived in Africa until the start of the Pleistocene epoch; paleontologists believe that a population of Australopithecus (3.5 mya) was the immediate progenitor of genus Homo, the line that eventually evolved (by the end of the Pleistocene) into Homo habilis (2.5 mya), Homo erectus (1.5-400,000 years ago), Neanderthal 125,000-20,000 years ago, Grimaldi (40,000-35,000 years ago), Cro-Magnon (25,000- 20,000) and our own species, *Homo sapiens* (200,000 years ago).

The details of the depigmentation process were triggered by environmental changes and the fossil lines trapped in the last Ice Age (1.75 million – 12,000 years ago) is where the mutations occurred in the hominid line (Diop, 1974; Finch; 1991). Table I suggests four different time periods for the mutation from dark to white skin. We can note that the Riss-Wurm interglacial between 120,000-75,000 years ago was a

144

period of significant warming which caused the glaciers to retreat, living the southern two-thirds of Europe habitable (Finch, 1991).

In these changing environments, lack of pigmentation or the absence of melanin in biological systems can affect a host of organisms from fish, reptiles and all types of mammals. It is not out of the ordinary to see white whales, gorillas, donkeys or big cats like Lions. Therefore, there is a mutation in the genetic line of many species that attributes to the change. Of prime importance, it is how other organisms of that species respond to the mutation that presents the controversy. It can be a negative response and lead to being ostracized or the organism can be embraced by the species without any disruptive experience. It is the curious mind of the human species that makes conjecture on this complex scientific alteration in history. Do you know any racist animals?

A real question is how to determine the formation of white skinned (Caucasoid) people with pale features in opposition to their genetic mothers and fathers? The human species is genetically similar across ethnic groups but phenotypically varied on multiple levels. For example, the number of DNA differences between races is tiny compared with the range of genetic diversity found within any single racial group. In humans, there is an interest to embrace the novelty, so in the past, the mixing of different genes through the phenotypic look may have been less problematic. People looking the same in appearance were then congregating together in a group collective. The result of this mixing is the variation we see today across the planet. In secure people, attractions come in all colors, shapes and sizes. For racist individuals, there is a fear of the other and the motivations toward the opposite appearance become devious.

Using Zebra fish, scientists have uncovered a wealth of information about the genetics of color variation. For example, this research has led to the discovery of a tiny genetic mutation that largely explains the first appearance of white skin in the human species tens of thousands of years ago. It is thought that the skin-whitening mutation occurred by chance in a single individual after the first exodus from Africa, when all people were brown-skinned (Weiss, 2005). Amazingly, the mutation involves a change of just one letter of DNA code out of the 3.1 billion letters in the human genome. When the genomes of four of the world's major racial groups were compared from a database, whites with northern and western European ancestry had a mutated version of the gene.

Furthermore, the research demonstrates that Asians owe their relatively light skin to different mutations. It is understood that light skin arose independently on different occasions in the human species and this is why we have facial features and other traits that are common hallmarks of Caucasian and Asian races (Weiss, 2005). White people are pale people who happen to be melanin-recessive, but they are not albinos. Humans of European descent have a mutation that does not allow a protein to permit melanin inside cells. It is interesting because the defect does not affect melanin deposition in other parts of the body like the hair and eyes, whose tints are under the control of other genes. Before we discuss albinism, this abnormal mutation is more than likely the genetic glitch that plays a role in the formation of "normal" white skin

Scientifically, albinism is a rare genetic condition associated with a variable hypopigmentation phenotype, which can affect the pigmentation of only the eyes or both the eyes and the skin/hair, resulting in ocular (OA) or oculocutaneous albinism (OCA), respectively. At least four forms of OCA and one of OA are known, associated with TYR (OCA1), OCA2 (OCA2), TYRP1 (OCA3), SLC45A2 (OCA4) and GPR143 (OA1) loci, respectively. A total of 15 genes are currently associated with various types of albinism. However, new genes have been recently described, associated with autosomal recessive oculocutaneous albinism with highly similar phenotypes but diverse molecular origin, indicating that there are likely to be more than 15 genes whose mutations will be associated with albinism (Martinez-Garcia and Montoliu, 2013).

Additionally, the rarest syndromic forms of albinism, affecting the normal function of other organs, can be grouped in Hermansky-Pudlak syndrome (HPS1-9) and the Chediak-Higashi syndrome (CHS1). The various disorders and syndromes highly indicate a whole host of defects that can occur when melanin is not working properly. Effects on the sensory systems cannot be overlooked.

Typically, albinism is characterized by decreased melanin synthesis, and sometimes this is associated with significant visual deficits owing to developmental changes during neurosensory retina development. As previously stated, all albinism is caused by genetic mutations in a group of diverse genes that affect enzymes, transporters, and G-protein coupled receptors. Interestingly, these genes are not expressed in the neurosensory retina. Further, regardless of the cause of albinism, all forms of albinism have the same retinal pathology, the extent of which is variable. There are currently seven known genes linked to the seven forms of ocular cutaneous albinism. These types of albinism are the most common, and result in changes to all pigmented tissues (hair, skin, eyes). The timeline for these changes is speculative, so let us review the fossil evidence, archaeological findings and biblical references pertaining to the appearance of melanin-recessive people (Table 2).

Diop (1974) and Finch (1991) have provided the most clear and logical written documentation on the origins of man. The phenomenal works of this teacher-student tandem have provided substantial research supporting the African origin of humanity and how melanin-recessive people were created. According to Finch (1991), we have established with the greatest degree of certainty from the anthropological record that humankind, from its earliest hominid ancestors four million years ago to its latest and most modern form as Homo sapiens sapiens 200,000 years ago, evolved entirely in Africa.

In sum, Finch states that the appearance of the Caucasian presents a compelling model of racial evolution. Through a complex of ecological factors figured in the process, the transition from black to white was the most decisive single event. The change of skin color, more than any other feature, put its stamp on the various races (Finch, 1991).

DATE	EVENT
400,000 BC	At the end of the Homo erectus presence on the planet, Homo erectus was found in many areas outside of Africa during the Ice Age. Dramatic changes in the weather could have triggered changes in physiology to adapt to the various climates outside of Africa. This supports the Multiregional Hypothesis that human variation occurred in different regions of the world.
200,000 BC	The "Mitochondrial Eve" in Africa is the established time period for the Homo sapienization process and the genetic origin of all humans on the planet today. Due to the intense radiation from the sun, a mutation in the genetic structure to make an albinoid appearance could have occurred on the continent of Africa or in a region outside of Africa. This supports the Out of Africa Hypothesis and the mitochondrial DNA scientific data.
40,000 BC	The presence of rickets in Grimaldi skeletons in southern France during the Ice Age indicates that these humans were forced to adapt to the sun-depleted environment by having dramatic alterations in skin pigmentation. This supports the view that a long-term cave experience is a sun-depleted environment can trigger a genetic mutation for the species to survive in harsh conditions.
5,000 BC	An interstellar or God-like influence could have manipulated genes and/or created a new variation of humans. This speculation supports both Sumerian texts and the Anunnaki as well as the biblical references in the Book of Enoch and his albino grandson, Noah, during this time period in history.

Table 2 - Approximate Dates for the Albinoid Presence and Melanin Mutation in Humans.

Merging Speculations and the Neutralization of Pigment Power

As previously presented, extraterrestrial origins and the fossil lines are speculative. By combining both perspectives, we are still engaged in speculation. However, there should be a point when science stands firm. What is factual is all humans carry genetic material from a woman who lived in Africa 200,000 years ago. The scientific evidence supports the Out of Africa Hypothesis and this is the mitochondrial Eve. In other words, Eve is the compliment to Adam related to the biblical Adam and Eve from the Garden of Eden. The Bible and the garden have nothing to do with science. The science is the fact that when the sperm and egg connect, it is only the woman's mitochondria that is biochemically found in the archaeological evidence. As scientist go back in time, the science proves that we are all African. Despite the location of humans all over the planet, the population of humans today all come from these melanin-dominant humans who first populated the planet Earth.

To present this more clearly, each person alive today can trace their ancestry to 10,000 people who lived 175,000 years ago (Perritano, 2019). We share 99.5% of our DNA with one another, so we are the human race. This is a powerful scientific fact because it explains how all of our DNA starts from a beginning point and all derivations that occur after still have a melanin-dominant base. Therefore, the appearance of lighter-skinned people (brown, white, yellow, red etc.) all have a genetic origin in blackness. As this point has been established in science, we can move further back in time.

Pangea and the connected continents, the presence of dinosaurs roaming the planet, and the formation of Earth's moon have nothing to do with humans today. Humans were not in existence during these historical events. When we speak of the hominid fossil lines leading up to 200,000 years ago, we have some clear demarcation points in history. Within those demarcation points, we can speculate on the time period for the mutation that caused melanin-dominant people to express a gene(s) to lighten skin and begin the transformation into other human forms (see Table 2).

149

In recorded history, there is a common belief that Sumer (5,000 - 4500 B.C.) is considered the oldest known civilization (Walker, 2006). Sumer was the southernmost region of ancient Mesopotamia (modern-day Iraq and Kuwait). This area is generally considered the cradle of civilization, and the name comes from Akkadian, the language of the north of Mesopotamia, and means "land of the civilized kings." The Sumerian called themselves "the black headed people" and their land, in cuneiform script, was simply "the land" or "the land of the black headed people" and, in the biblical Book of Genesis, Sumer is known as Shinar.

Molefi K. Asante has been a foundational figure in the African-centered movement over his career, and he shares the view that many contemporary Eurocentric historians and intellectuals use Mesopotamia as the cradle of civilization to negate a melanin-dominant presence at the beginning of recorded history. Mesopotamia does not figure in ancient civilizations, either concretely or philosophically, at the same level as ancient Egypt (Asante, 1999). Even were one to take evidence from ancient Egyptian, Hebrew, Greek, and Ethiopian people, one would find that the Nile Valley of Africa rather than the Tigris-Euphrates Valley in the "Middle East" was considered the most ancient cradle of human civilizations. Asante understands that Mesopotamia was a significant civilization, but he believes it is advanced as a sort of contemporary counterpoint to the African origin of civilization.

It should not be difficult to realize that black people were all over the world establishing various civilized experiences and influencing other cultures in different regions of the planet. Therefore, we can historically observe many parallels in recorded history. For example, when we analyze Dogon systems of knowledge, we see links to the ancient Nile Valley and ancient Sumer (Finch, 1998). Finch reveals that some features of the Dogon Venusian calendar resonate in ancient Sumer and Chaldea. Also, the eight Dogon nommo are all fish-beings in the style of Oannes of the Fertile Crescent. Ostensibly, there is a Cushite or melanin-dominant influence from the beginning and the cultural comparisons show an unmistakably Nilotic provenance for many aspects of Dogon culture and ancient Sumer.

In one quick swipe of a writing utensil or slash and burn of documents, history can be tragically rewritten and falsified (Figure 17).

Fig. 17 - Statue of Ptah Hotep with face damaged.

It is odd for some staunch white intellectuals to arrogantly claim that Afrocentric scholars are teaching myths and not real history, when it is not black people who deliberately destroyed information to wipe out historical texts and documentation. After the introduction of literature and writing to a people who had been formerly living like savages, history is replete with displays of aggression on true history by melanin-recessive cultural groups. The blatant theft by Greeks around 640 B.C., Alexander and Aristotle pillaging ancient African libraries in 332 B.C., the burning of the Library of Alexandria in 44 B.C., the Spanish Inquisitions since 1250 A.D. and the expulsion of the Moriscos in 1609 A.D. are all examples of a white fragile mindset gone wild. Cultural deconstruction and book-burning were carried out on an unheard-of scale, and the destructive effects of this book-burning on all branches of learning is beyond comprehension (Finch, 1998). In a chapter on African Navigation and Cartography in his book, Finch (1998) states that, "Gradually, the achievements of ancient world science became progressively a dimmer memory. The surviving body of Greek literature that managed to be passed down lent itself to the deepening perception that such learning as did exist in earlier times was Greek and

151

Greek alone. In truth, this Greek learning was but a pale after-reflection of a sun that had already set." p. 233.

As an individual, comprehend that this historical area is rich in the history of melanin-dominant people, and reflect on the impact of the devastation of the U.S. led imperialistic war against Saddam Hussein and the Iraqi people. In 1990, Saddam Hussein was given international permission to attack Kuwait for slant drilling into Iraq's oil supply, and he was set up for an unwinnable war. It was said he had weapons of mass destruction. Years later, we can see the travesty of the fake war against Saddam Hussein and the Iraqi people. With all those bombs and explosions, think about the ancient history that was destroyed by the massive assault on the land. It was a staged hit on history, and evidence for early civilizations was destroyed in the bombings. We know there were no weapons of mass destruction, so what we had was mass deception.

In the beginning of this decade of 2020, we have fake people and fake leaders manufacturing fake wars and leading us into real consequences for real people. "Provoke and stoke" seems to be the World War III motto that can only benefit the rich. A pusillanimous drone attack ordered by #45 to kill an Iranian leader (i.e., Qassem Soleimani) has spun the world even more out of control. For #45 to even utter the destruction of Iran's cultural heritage sites can further alter world history in Iraq's neighbor, Iran.

Even today there is confusion about the pigmentation of the original inhabitants of Sumer, their origin and where they came from in history. If we know there were massive monuments found in other parts of the world before 5,000 BC, we cannot close our minds to the existence of civilizations before Sumer. This cradle of civilization was really the northern region of Africa, and it was given credit for the first due to the cuneiform system of writing. We know the language and writing was not Indo-European and not Semitic. However, this form of writing was later adopted by their Semitic conquerors who more than likely rewrote the actual history of the era. When you control the narrative, you control the perception of the story. This is where merging speculations can neutralize pigment power.

152

The region of Sumer was long thought to have been first inhabited around 4500 BCE. This date has been contested in recent years because human activity in the area began much earlier. When written and interpreted by western, Eurocentric researchers, it is acknowledged that the first settlers were not Sumerians but a people of "unknown" origin. In contrast, Houston (1985) writes in her classic book, *Wonderful Ethiopians of the Ancient Cushite Empire*, that the original people were black. According to Houston, the foundations of ancient Chaldea were laid as early as those of Egypt. The Sanskrit books of India called Chaldea one of the divisions of Cusha-Dwipa, the first organized government of the world.

Houston documented this in 1926 that all the earliest traditions of Chaldea center about Belus and Nimrod, and Nimrod was the son of Cush. She continues, "Babylon had two elements in her population in the beginning. The northern Accadians and the southern Sumerians were both Cushites. The finds of recent explorations in the Mesopotamian valley reveal that these ancient inhabitants were black, with cranial formation of Ethiopians." Therefore, no more time will be spent on the appearance, even though the artwork displayed from this era has bearded European-like images.

Not many African-centered scholars spend much time attempting to explain these Nordic looking fish-like images of white men with bags on one hand and a pinecone in another. How authentic they might be is questioned because we know Jewish invaders came into Sumer later in time to rewrite the history. The story of cursed people began with Jewish mythology and this false interpretation of history has been adopted by other doctrines like the Mormon religion. Even more fantastical is the association of these white figures to be from another planet. If true, then here is where the embellishment confuses future readers.

For example, the Bible makes reference to Nephilim or fallen angels that have come upon Earth from the heavens. True or not true, we now have multiple reference points to an astrobiological connection to our existence from ancient texts. Oannes is supposed to be the name of the bearded white man (Figure 18), and he is supposed to be the civilizing hero from before the flood.

**Fig. 18 - Ancient Anunnaki image of
a fish-like man from Sumer.**

These are the same fishman-like images in the Dogon cosmology. Occultist suggest the bag in the left hand contained the knowledge of how to rebuild a civilization and the pinecone represented the pineal gland, for spiritual wisdom perhaps. Onannes is from celestial people known as the Anunnaki, and this conceptualization of extraterrestrial beings during this time period in history can be found in the Book of Enoch and the Bible. These figures are representations of the people who manipulated human genes 445,000 years ago (Sitchen, 1991; Evans, 2016) or came to assist with human development again around 3,000 B.C. in Mesopotamia. The Eurocentric storyline is the Anunnaki traveled from their planet Nibiru to extract gold from Earth using Homo sapiens as slaves. It is believed the Anunnaki genetically manipulated apes to make humans as their slaves on planet Earth (Figure 19).

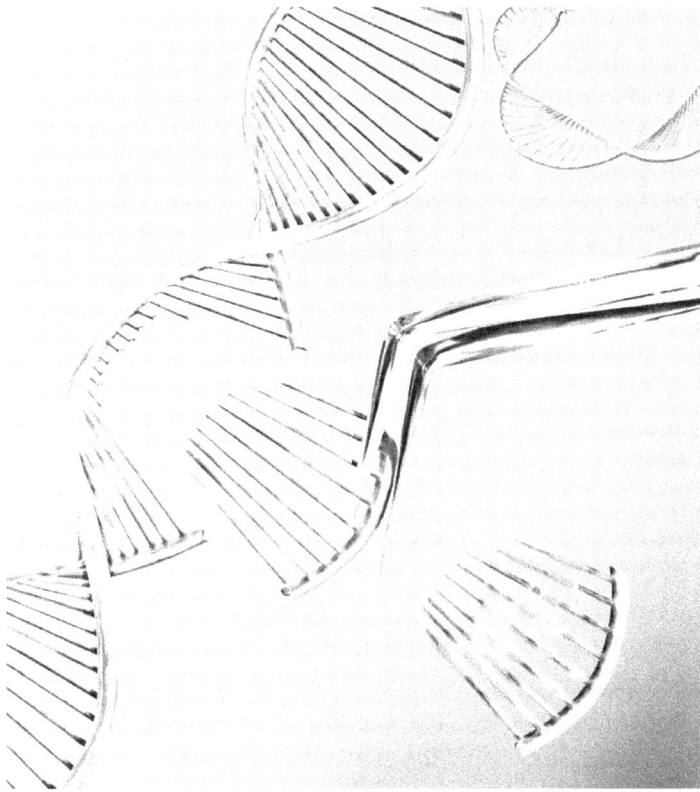

Fig. 19 - A sample model of a delicate DNA splice that can only be performed with microscopic technology.

Houston (1985) makes a more logical conclusion about these images and the fish-garbed figure. The people there in Mesopotamia were living like animals without any order of government. Nimrod saw the fruitfulness of the land and set up a foundation for a civilized society. According to Houston, "He sent to them from the sea, a fearful fish by the name of Onan. This was a ship which appeared to these barbarous people as a great fish. Its image half man and half fish is still preserved. It represents men who came to these untutored people by water. These primitive people of the Mesopotamian valley had not yet conquered the sea and this happening was perpetuated from generation to generation as they were first impressed." Whether we believe those images were the original people of that land, what we do know is that there were conquerors who rewrote the history. Just like many African/Black images have been erased from our view of ancient Egypt,

155

it is understood how this could happen in Sumer (with the original inhabitants), India (Indus Valley Civilizations), Central America (Olmec-Mayan Civilizations) as well as southern Europe with the Moors (711-1492 A.D.).

On this topic of manipulating history, there are points to connect here. The first point is the Nicene Conference in 325 A.D. organized by Constantine the Roman Emperor. In the manipulation of Christianity, this is where certain books were left out of the biblical scriptures. Secondly, although the Book of Enoch was considered at the conference (Dudley, 1992), it was left out of the Bible. It was interesting that this book (Laurence, 1995) was purported to document events that occurred around 5000 B.C.

The Book of Enoch is a very strange document, purporting to be a vision of the future cataclysm of the flood, and why it was unleashed upon the world (Hancock, 2015). Towards the end of the book in Chapter CV. Enoch's son named Mathusala tells the experience of his son Lamech and his wife (no name??) having a son who brought trouble to his psyche. In verse 2, *She became pregnant by him, and brought forth a child, the flesh of which was white as snow, and red as a rose; the hair of whose head was white like wool, and long; and whose eyes were beautiful. When he opened them, he illuminated all the house, like the sun; the whole house abounded with light.* It is quite obvious this was an albinoid human characterization that was being described to these melanin-dominant parents. They could not understand for they were black, and their son came out white and lacking melanin.

In verse 3, *And when he was taken from the hand of the midwife, opening also his mouth, he spoke to the Lord of righteousness. Then Lamech his father was afraid of him; and flying away came to his own father Mathusala, and said, I have begotten a son, unlike to other children (meaning a changed son). He is not human; but, resembling the offspring of the angels of heaven, is of a different nature from ours, being altogether unlike to us.* Of course, the child was human, but the father could not psychologically deal with the albino appearance. Later in the chapter it is revealed that the son was named Noah. Yes, the same Noah associated with the flood, the Ark, his three sons, and the "cursed" Ham. The Bible has stories upon stories.

156

During this time period described by Enoch, it was several centuries before the Christian era. Moon worship, sun worship and star-worshippers were the movements of that day. In several regions of the Earth, massive structures were developed on astronomical alignments. In Enoch's stories from this time, fallen angels and "watchers" from above were in existence. The current Bible refers to these entities as Nephilim. Whether we believe the shamanistic-type vision or dismiss the possibility that Enoch was describing actual events, we know we are not alone in this grand universe. Miles away from Sumer (i.e., Iraq) is Turkey and Egypt. Almost 5,000 years before Sumer we have massive structures like the immense stone block in the quarry of Baalbek and the Hor-em-Akhet in Kemet that appear to have astronomical connections. During this early period in history, even Stonhenge in England (3000 B.C.) is associated with astronomical alignments. Therefore, much had occurred in history before and after Sumer, and it is unexplained as to how these monumental structures were made. What technology existed during that time period to eventually lead up to the construction of the enormous pyramids all over the world? How was this technology wiped out of our memory?

Ice Age Evidence

At the end of the last Ice Age, the sudden onset and equally sudden termination of the mysterious epoch known as the Younger Dryas changed the course of history for humanity. The Younger Dryas was a major and abrupt change of the world climate which happened from roughly 12,900 to around 11,700 before present time. This means that the event took place about 13,000 years ago and lasted for about 1,300 years. Due to the Ice Age experience, the thaw, and the cataclysmic events, the human body naturally responded to the circumstances. The mutation in the melanin gene was subtle and we can speculate that is what the biblical character known as Noah experienced during a significant period in history. In Table 2, there are four different periods in history I would consider as a match for the biblical, archaeological and genetic evidence for when the change occurred in Homo sapiens.

An excellent synopsis of this recent ice age period is provided by Hancok (2015). In his assessment, somewhere around 12,800 years ago, after more than two thousand years of uninterrupted global warming, a flood of icy meltwater entered the North Atlantic Sea so suddenly, and in such quantities, that it disrupted ocean circulation and, subsequently, coastal lines were swallowed up. In the same geological instant that the meltwater flood was unleashed, global temperatures plummeted, and the world's climate underwent a reversal from the balmy 2,000 year long "summer" that had begun about 15,000 years ago to a savage and icy global winter. At the end of this tumultuous period, around 11,600 years ago, the freeze suddenly ended, global temperatures soared, and the remnant ice caps collapsed, shedding their residual water burden into the world's oceans which rose dramatically to close to today's level. As a result of these dramatic climate changes, getting stuck in the ice would eventually change the texture of humanity inside and out. Every region of the world was impacted and some more than other locations on the planet. The dramatic impact could be the immediate loss of knowledge, science and technology.

We have historical evidence that ancient people on the continent of Africa were smelting iron ore in 43,000 B.C. There is recent evidence of the oldest cave paintings yet found on the Island of Sulawesi near Indonesia around the same time period (Guarino, 2019). In modern times, however, video evidence has shown how a tsunami could wipe out a coastal community in a matter of seconds (e.g., 2004 Indian Ocean earthquake and tsunami in Sumatra). All historical evidence could be changed forever and wiped from our memory banks with a single cataclysmic event. Due to the global warming crisis we are now experiencing, there are some islands (e,g,, Fiji) that are slowly drifting underwater and major land masses (e.g., Australia) being destroyed by fire.

Horace Butler (2009) reiterated that there are lost civilizations we have yet to officially document for the historical record. According to Butler, he can document civilizations that have existed 22,000 years ago and how a melanin-dominant people civilized the Americas thousands of years ago. In fact, Butler provides a thesis that the Egypt in the Bible was really an Egypt in the Americas. He uncovers ancient writings supporting his claim that a civil war in Central and South America had forced 'Egyptians' to flee from the Americas back across the Ocean to Africa. They escaped to the eastside of Africa, where they built a second Egypt (Butler, 2009). It is interesting because the historical record on the walls in Kemet do reference places like Punt (Somalia) and the Ethiopian region of Africa as their place of origin.

In the comfort of your home, you may have difficulty imagining a need to immediately relocate. However, in ancient times, it was an experience that had to be confronted. As a matter of fact, it is a devastating experience that must be confronted today, but many industrialized nations, led by megalomaniac leaders of corporations, are presenting an illusion that there is no global weather crisis to worry about, and this is not intelligence.

Historically, people during ancient times were forced to navigate the world. They were intelligent. One does not wait for the impending doom with no plan of action. It is wiser to create a reality to increase your longevity and your historical presence on the planet. Intelligent humans should be beyond the conceptual incarceration displayed by racist scientists to think melanin-dominant people all over the world were waiting around for "white saviors" to tell them how to circumnavigate the planet. The presence of African people all over the planet clearly negates a cultural isolationist point of view.

In Finch's (1998) detailed chapter on navigation, there is overwhelming evidence for research supporting claims made by Butler. There is modern evidence from *Kon Tiki* voyages demonstrating the capability of ancient Peruvians completing westward voyages across the Pacific and the geographic range in reed boats was from Africa to the Andes. The Nilotic and the Andean people have many cultural parallels and boats, and not spaceships, made the connections over historical time. Finch goes on to explain from the research he has reviewed, "Even in the modern era, users of reed boats could be found in Ethiopia, Chad, the southern Sudan, Sardinia, Morocco, Mesopotamia and Bolivia. The earliest reed boats are depicted on the Egyptian monuments and vases and go back to pre-dynastic times." p. 212.

Furthermore, Butler highlights Peru as a citadel for ancient African explorers, so it is intriguing to see the development of Machu Picchu (1450 A.D.) high in altitude. Apparently, they remembered their African unconscious survival mechanisms and created an impeccable city that will last in perpetuity. They built this massive complex and historical wonder of the world high in the mountains to ensure a flood would not decimate the population in that region. I surmise the people choose the region far removed from the coastline to avoid being swept away under water in contrast to other lost civilizations that are currently under water. The question to ask is, "When will we get swallowed by the sea under the current global warming crisis?" (Figure 20).

Fig. 20 - Inside a glacier in Greenland and an increasingly expanding river impacted by the global warming crisis.

Today, the corporate controlled news media will not show melanin-dominant people being displaced by this climate crisis and this is tragic. The media ignores the reality that in places like the Melanesian Islands and Australia, there are African-looking people. Sadly, many islands in the Pacific Ocean are sinking under water, tsunamis are causing displacement in Indonesia, and the fires in Australia and North America leave people with nowhere to go. In ancient times, circumnavigation was one of the solutions to survive these cataclysmic events. Therefore, the reader should now visualize how and why the African presence has expanded throughout the world. In this expansion, the genetics of the people changed to adapt to the environments they relocated.

Science can prove that environmental conditions triggered a change in the genes for pigmentation (i.e., melanin). On a cosmic level, melanin is ubiquitous, and it is the astrobiological source to our African unconscious (Bynum, 1999). It is problematic to attribute a God cursing a people as a reason why we have color variation in the human race. Being cursed from an unknowing entity makes no logical sense. Do we actually think a righteous God would favor one group of people over another? No, cut the nonsense about a "chosen" people and a cursed people. If anything, it is a dismissal of evil and wicked people who the scriptures focus on as the type of people God would choose to eliminate.

Beyond God, some people have attributed the genetic manipulation of the human species by interstellar explorers (Sitchin, 1991). In other words, there are people who believe in the presence of the Anunnaki (white European looking people) from another planet who created the human species from lower hominids. The suggested dates could be 445,000 years ago or 3,000 years ago when these visitors to planet Earth came from above. Although it is expressed that human beings are made up of star dust and we are a part of the cosmos, all effort on this planet was created, developed, built from the mind and capabilities of humans walking this Earth. No matter how we view our origins, what humans have done, humans can do.

History can be deliberately manipulated and/or destroyed by the ravages of nature. In either case, we cannot get confused about the bible, archaeology and general history. If we add the information age and the internet, we have another layer of distorted history that can lead people astray. As we conclude, take note that historical amnesia can wipe you from the face of the Earth, so read, study, and put your King "whatever your name is" Version together. None of us were there back in time, but we are present today to determine our destiny here and now.

CHAPTER 12

PIGMENT POWERED DEFENSE SYSTEM

Curiosity killed the cat; Naivety depopulated the people with a bat.

Interstellar Germs

In the last chapter, we speculated on contact with beings or entities from other planets or galaxies. One thought to consider is how a foreign agent from another world could impact our biological system here on Earth. It could be advantageous if it contained the DNA markers of the human genome. In contrast, a substance from another planetary existence could be a potentially detrimental contagion that our bodies could not have a defense against. The danger from a foreign astrobiological element entering our atmosphere could be ultimately fatal to all life on earth.

It is a fact that we are not immune to any particular antigen. Our immunological system is actually a defense mechanism system, and the human body can create antibodies to fight the foreign agent. If this foreign agent contained a genetic element that was unknown to our natural biological system, the impact could wreak havoc on the human species. Whether the foreign agent haphazardly arrived on a meteorite or if it was manipulated by entities to deliberately affect people, humans could be wiped out. Therefore, it is important to have responsible leadership at the helm if we ever made contact with a foreign agent which has never been experienced.

Dark matter or blackness pervades the Universe. On an interplanetary level, melanin is found throughout the cosmos. Melanin is a complex biopolymer that can absorb biological agents, and the power of pigment may be the first line of defense against an interstellar contagion. If black fungi can be found forming in abandoned highly radioactive nuclear power plants, then melanin is a universal molecule with unimaginable qualities for survival and defense.

Melanin is a major defense molecule in invertebrates but its role in mammalian immunity remains less explored. In contrast, several recent studies have highlighted the emerging innate immune activities of human melanin-producing cells which can sense and respond to bacterial and viral infections. Indeed, the skin is a major portal of entry for pathogens such as arboviruses (Chikungunya, Dengue) and bacteria (mycobacterium leprae, Leptospira spirochetes). Melanocytes of the epidermis could contribute to the phagocytosis of these invading pathogens and to present antigens to competent immune cells. Melanocytes are known to produce key cytokines such as IL-1β, IL6 and TNF-α as well as chemokines. These molecules will subsequently alert macrophages, neutrophils, fibroblasts and keratinocytes through unique crosstalk mechanisms (Gasque and Jaffar-Bandje, 2015). If these events can happen in the skin, melanin can also provide immunoenhancing mechanisms to fight viruses internally, externally as well as foreign.

Zoonotic Diseases

An example of agents that can be truly foreign and detrimental are zoonotic viruses. Zoonosis is an infectious disease caused by bacteria, viruses or particles that spread from non-human animals to humans. This is important to discuss because of the world-wide pandemic of the coronavirus known as COVID-19 (or SARS-CoV-2).

Scientists have always worried about a pandemic since the days of the Bubonic plague in ancient times, the 1918 Spanish Flu outbreak, and the 2009 H1N1 Flu epidemic. In addition, the scary SARS (Severe Acute Respiratory Syndrome) and MERS (Middle East Respiratory Syndrome) epidemics decades ago continued to demonstrate that the world could be in trouble from emerging viruses. In nature, the jackpot in flu evolution occurs when two different types of viruses get into an animal cell at the same time.

Intense investigations are going on to find the cause of the current SARS-CoV-2 pandemic. Recent scientific evidence from the Nature Medicine journal indicates that the new coronavirus did not escape from a lab (Anderson et al., 2020). This research team from The Scripps Research Institute compared the SARS-CoV-2 genome and seven other coronaviruses known to infect humans: SARS, MERS, and SARS-CoV-2, HKU1, NL63, OC43 and 229E. The first three cause severe disease, and the latter three cause mild symptoms.

This data has come out very quickly to indicate the SARS-CoV-2 is not a laboratory construct or a purposely manipulated virus. The key evidence is on the spike proteins that protrude from the surface of the virus. The coronavirus uses these spikes to grab the outer walls of its host's cells and then enter those cells.

Another source of evidence presented by the researchers is that the overall molecular structure of this virus is distinct from the known coronaviruses found in bats and pangolins that had been little studied and never known to cause humans any harm. In the case of SARS-CoV-2, the researchers provide their suggestions on the origin of this novel virus that has disrupted our daily lives.

These researchers believe the common zoonoic jump from mammals to humans occurred in this current pandemic. For example, we contracted SARS from civets, MERS from camels, and H1N1 from birds and pigs. Therefore, the researchers suggest that an animal like a bat transmitted the virus to another intermediate animal (possibly a pangolin) that brought the virus to humans. In the open markets of Wuhan, China, many live animals are sold and eaten. The exposure of the contagion from these animals has obviously wreaked havoc on humanity.

Another possibility is the pathogenic features would have evolved only after the virus jumped from its animal host to humans. Some coronaviruses that originated in pangolins have a "hooked structure" that makes it easy to enter human cells. The virus is more capable to spread between people because once inside the human host, the virus could have evolved to have a stealth feature – a cleavage site that allows it to easily penetrate human cells.

Beyond the cleavage site, the hook is the key. The new virus attaches to a specific receptor on respiratory cells called angiotensin-converting enzyme 2, or ACE2. To infect a human host, viruses must be able to gain entry into individual human cells. They use these cells machinery to produce copies of themselves, which then spill out and spread to new cells. The key to this puzzle is the interaction with the spiked protein (S-protein) on the virus with the ACE2 protein on the human cell.

Researchers at Westlake University in Hanzhou, China breakdown the process with the analogy of a robber breaking into a house (see Pappas, 2020). COVID-19 is the robber, and the house is the human body. The doorknob to get into the house is ACE2. The S-protein on the virus grabs onto the ACE2 protein and easily enters the body. It is like a master key into your respiratory system. This is devastating because the easy entry is due to the sticky nature of these components and the viral particle is more likely to enter a cell and the stickiness makes it highly contagious.

Blood Pressure Issues: The Key Link

Let us break down this science even further than these researchers are presenting. Angiotensin is a natural chemical that can cause vasoconstriction and shrink blood vessels. The precursor of angiotensin is angiotensinogen which is synthesized in the liver. ACE2 is actually a hormone in the human body, and ACE2 inhibitors are synthetic drugs that keep the blood vessels open. There are over 100 different types of blood pressure medicines, and ACE2 Inhibitors are an effective blood pressure medicine for many people. However, blood pressure medications can be ethnic specific (Iniesta, et al., 2019), so **melanin-dominant people, GET OFF ACE2 inhibitors** to lessen your susceptibility to COVID-19. Research has demonstrated that ACE2 inhibitors can lower the immune response, and they are not useful for monotherapy for melanin-dominant people (Helmer, Slater, and Smithgall, 2018). In fact, hypertension in African Americans is more responsive to monotherapy with diuretics and long-acting calcium channel blockers (Johnson, 1999).

All over the world, older people are dying from COVID-19 more easily because some people are taking multiple types of synthetic medications. I cannot say that ACE2 inhibitors are the main reason, but we know that Italy and Spain were hit hard due to an older population.

In combination with the Italian diet of pasta and carbohydrates that can feed the virus, the older Italians have suffered tremendously. The medical establishment becomes unreliable to resolve this pandemic because diet, nutrition and exercise are seldom addressed as an important factor. Instead, more drugs, other medications and vaccinations are the running theme.

According to a recent news report (Perper, 2020), Italy's population is the second oldest in the world, after Japan. During the height of the pandemic, Italy had suffered the worst outbreak outside of China. However, the USA has taken over the number of deaths per capita, especially in the black community. In Australia, only 15 per cent of Australia's population is aged 65 or above and that country is not as devastated (Bain, 2020). Sadly, more than 99 per cent of people who died from coronavirus in Italy had pre-existing illnesses, according to the country's national health authority. The most common problems included high blood pressure (76%), diabetes (35 %) and heart disease (33 %). The statistics suggest thus far that 76 % of the patients who died in Italy had high blood pressure. In this news article by Bain (2020), there is a random opinion from Alister McNeish, an Associate Professor in Cardiovascular Pharmacology at the United Kingdom's University of Reading. He said there is no direct evidence that medications to treat high blood pressure, such as ACE inhibitors, lead to increased mortality. The disturbing recommendation from this melanin-recessive researcher from England is that many patients with high blood pressure in Italy already have an overactive renin-angiotensin-aldosterone system which could mean an increased ACE2 expression regardless of treatment. As a result, his recommendation is that patients remain on their medication.

As previously stated, research indicates that melanin-dominant people do not respond well to ACE2 inhibitors. They should not, therefore, take that type of advice provided by the researcher in the United Kingdom. Even though they are prescribed, it is not recommended for melanin-recessive people to take ACE2 inhibitors, especially during this coronavirus pandemic. Request a change in medication and save yourself from unwanted health disturbances. In contrast, many Black people on blood pressure medicine respond better to calcium channel blockers. Deeply reflect on this since it is a matter of life and death.

It is well established that calcium ions are critical for nervous system functioning and muscle activity. When a nerve impulse depolarizes a presynaptic membrane, there is a calcium ion influx. In addition, calcium ions are necessary for the contraction of actomyosin and neurotransmitter release (Feldman and Quenzer, 1984). Electromagnetic frequencies or Wi-Fi exposure causes calcium overload (Pall, 2018), so analyze the problems experienced by people who have been attacked by the virus. Beyond the severe respiratory issues, the heart and the cardiovascular system and the kidneys are impacted in COVID-19 patients.

In other words, this new 5G technology coinciding with the COVID-19 virus can potentially interact with the bodies of people on high blood pressure medications. All the scientific evidence from countries throughout the world indicate that individuals with high blood pressure, diabetes and obesity are at a greater risk to succumb to the COVID-19 virus. Some blood pressure medicines are calcium channel blockers. Therefore, if you really need medication, you will probably have a better survival rate with calcium channel blocking medication. With this mass rollout of 5G technology throughout the planet, it is affecting biological systems in adverse ways. For example, if the calcium channel blocking drug prevents calcium overload, then it might be a more useful medication during this unprecedented viral pandemic when 5G technology has shown to cause voltage-gated calcium channel activation (Pall, 2018).

In summary, pre-existing conditions are highly problematic when attempting to effectively treat a patient. On numerous occasions, it is often too late for many people to get proper care because of a lifestyle of poor health choices and a campaign of discriminatory practices against non-white communities. The evidence pertaining to who is dying more rapidly from COVID-19 in the USA is tragic, and now one can objectively see a "Plandemic" unfolding. The immune system must be built up and not weakened. The presence of melanin can be protective, but it can be a double-edge sword as it absorbs dangerous outside energy. The presence of melanin can raise higher levels of interferon in the system to fight viruses, but pollution and the intake of synthetic substances can lower your immunity. As a biomedical researcher, I recommend that melanin-dominant people should avoid drugs like ACE2 inhibitors because it will make the body more susceptible to the COVID-19 virus due to easy entry into human cells.

To treat COVID-19, current vaccine researchers are investigating this viral spike protein for ACE2 as a promising target as well as promoting interferon medications [Interferon Alpha-2B Recombinant (IFNrec)] stemming from research in Cuba (O'Connor, 2020).

Internal Threats to Humanity

Even though the aforementioned researchers from Scripps Research indicate this novel virus was not created in a lab, we must not be naïve and doubt that the virus could have been naturally uncovered, manipulated and released on purpose with an ulterior motive. In 1982, Robert Harris and Jeremy Paxman wrote *A Higher Form of Killing* to document evidence on the secret story of chemical and biological warfare. The chapter on The Search for the Patriotic Germ was very revealing. Interestingly, there was no mention of AIDS or HIV during this publication, but there were a host of other diseases that were investigated before the AIDS crisis hit humanity.

In excerpts from the 1982 book, the authors state:

> "Diseases might be sprayed into the air from a ship or aircraft, and allowed to drift across the country. To discover whether such attacks, feasible in theory, were practical propositions, the British, Canadian and Americans collaborated in a succession of experiments. After preliminary meteorological research to discover how clouds of bacteria might behave at altitude, they began a series of mock attacks." p. 155.

Currently, we see how severe the COVID-19 has impacted New York City, the California Bay area, San Francisco, and many other big cities in the USA. The authors revealed:

> "In 1951, American Naval personnel deliberately contaminated ten wooden boxes with *Serratia marcescens, Bacillus globigii* and *Aspergillus fumigatus* before they were shipped from a supply depot in Pennsylvania to the navy base in Norfolk, Virginia. The tests were designed to establish how easily disease might be spread among the people employed to handle boxes at the supply depot. Of the three infectious bacteria, *Aspergillus fumigatus* had been specifically chosen because black workers at the base would be particularly susceptible to it." p. 152.

The use of aircraft is more effective than a flying bat to spread diseases. The authors state:

> "Beginning in the spring of 1957 RAF aircraft were regularly dispatched on missions around the British coast. From specially constructed tanks slung below the planes they poured out zinc cadmium sulphide, a chemical easily detected, even in minute quantities, in the atmosphere. Monitoring stations were established across the British Isles, where Porton scientists assessed the quantity of the chemical in the

air. By the autumn of 1959, when the experiments were completed, almost the whole country had been sprayed with the chemical." p. 158.

We know German doctors crossed lines after WWII to join the United States. The bioweapon research knowledge from these Nazi doctors was used, and according to the authors:

"But the main objective was the development of a weapon to kill people…It should be a disease against which there is no natural immunity. It should be highly infectious, and yet the enemy should not be able to produce a vaccine against it to be able to cure the disease with the medical facilities available to him. And from a military point of view, it should be a disease which was easy to reproduce itself outside the laboratory." pp. 160-161.

There is much more to quote relevant to the current story in world history, and the connection to campaigns against China go back to the 1960s when research was being conducted. Again, this information is decades old and according to the authors:

"The secret spraying carried out in the United States, Britain and Canada had provided critical information about how thick a cloud of bacteria needed to be to spread disease successfully. Experiments at Fort Detrick and Porton Down had shown how long microorganisms would live while floating in the air. Tests on animals had provided invaluable information about how large the individual particles needed to be to break through the body's natural defences." p. 164.

Armed with this information, Chemical Corps generals began to imagine astonishing biowarfare campaigns against Communist China sixty years ago. It was known back then that China is subject to polar outbreaks and the airflow from Siberia can cause agents to easily spread.

All of this previous research is documented, and it has set the stage for a play book that is read and practiced by the 1% who control the wealth of the world. For example, on May 5, 2009, some of America's leading billionaires met in a private Manhattan home just a week before the annual meeting of the secretive Bilderbergers (Marrs, 2015). It would be interesting to know if the meeting was organized at the home of the deceased Jeffery Epstein.

These meetings were closed but it is not as though the agenda could be put on full blast to the public. Our minds are conditioned to think people with big money are kind, benevolent and philanthropic. There is a reason why they have accumulated mass wealth, and more than often, the deeds behind getting the money are dastardly. Also, a billionaire is not interested in sharing the wealth, they are concerned about keeping the wealth and this is an internal threat to humanity because they will do whatever is necessary to keep a stranglehold on their wealth.

Marrs (2015) informs us the attendees reportedly included Bill Gates, David Rockefeller, Jr., Warren Buffett, George Soros, Micheal Bloomberg, Ted Turner and Oprah Winfrey. They were there to discuss depopulation strategies instigated by Gates whose agenda advocates that human overpopulation was a priority concern. Note that Gates has invested nearly $71 million into Planned Parenthood over the years. A consensus was probably reached that they would back a strategy in which population growth would be tackled as a potentially disastrous environmental, social and industrial threat.

Let us fast-forward the play book to the current COVID-19 pandemic. On October 18, 2019, there was a Global Pandemic Exercise conducted at The Johns Hopkins Center for Health Security in partnership with the World Economic Forum and the Bill and Melinda Gates Foundation. The event was called Event 201 as a partnership to respond to a "potential" severe pandemic to diminish large-scale economic and social consequences.

If you cannot see the connections to the same people and the same plans, then your mind has been mentally incarcerated. The author of *Emerging Viruses*, Len Horowitz (1996), reminds us about these repeats in history. To make it plain, the millions of Holocaust victims were told they were going into showers for public health and to be disinfected. The gullible were tricked to believe something different, and the gas chambers were apparently the mission of Nazi doctors. After creating the COVID-19 pandemic, they now will create the solution and profit from mass vaccinations. The agenda for population control has been written and the viral pandemic movies have been made.

Silent Ethnic Weapons

Despite evidence that fears of population growth are overblown, the globalist seeking population reduction have continued their systematic elimination of huge numbers of people (Marrs, 2015). Following the 2009 outbreak of swine flu, it was found that the strain contained a combination of genes from swine, bird, and human influenza viruses. Because this virus could not be contracted by eating pork or pork products, researchers suspected swine flu was manufactured by humans. Even back in 2009, it was believed that this was one of the venues used to reduce the human population by the global elite, who have long supported eugenics. In fact, medicine is a perfect instrument to use for this strategy because it is easy to make people believe that draconian measures are necessary to protect the public health, especially if the public is primed with the right climate of fear (Banks, 2010).

It is virtually impossible to run from something dangerous that you cannot see. What is generated in the body is fear and fear can make the body sick. Just like there is a placebo effect and people can improve health from a belief in their mind, think about the reverse scenario. The heightened reactions in the body set up a certain frequency that can be manipulated from an external level. Think for a moment about a singer who can hit a high frequency note and crack a glass. Likewise, the body can also be impacted by frequencies that can leave cells in disarray.

Scientists may one day be able to destroy viruses in the same way that opera singers presumably shatter wine glasses. New research mathematically determined the frequencies at which simple viruses could be shaken to death. "The capsid of a virus is something like the shell of a turtle," said physicist Otto Sankey of Arizona State University. "If the shell can be compromised [by mechanical vibrations], the virus can be inactivated" (Schirber, 2008). The experimental evidence has shown that laser pulses tuned to the right frequency can kill certain viruses. However, locating these so-called resonant frequencies is a bit of trial and error.

As the technology has improved, probably in secrecy, the possibility to use frequencies to destroy viruses could save money for the patient. Although this would save money for the patient, the medical-industrial-pharmaceutical complex would greatly desire a drug solution to make money. In a diabolical manner, the use of this unique technique could also activate viruses to flourish.

For this reason, there has been a theory that the new distribution of Fifth Generation Technology (5G) on a world-wide scale has fueled the cells for COVID-19 to flourish. For the virus to impact every region of the Earth seems like this could be a serious theory to consider how the pathogen has spread. Africa is less digitally connected and the incidence on the continent of Africa is much lower than other regions of the earth. Many cases in Africa have been due to travelers from abroad who brought the virus to a location where massive numbers of melanin-dominant people live. South Africa has large cases of COVID-19, but these stem from white, melanin-recessive South Africans returning home from traveling abroad. Chinese are also invading the continent for work projects in West and East Africa, so there are other underlying entry points into Africa. However, scientists are bewildered as to why it has not decimated the African continent.

Another silent weapon is spraying in the air. Think about chemtrails in the sky and the chemicals being sprayed. Chemtrails have been going on for years, so this is another theory that cannot be neglected. In this time of global crisis, there are still military planes flying high in the sky well above altitudes for commercial airlines. These aircraft could not be commercial planes because the aircraft are spraying chemicals in patterns that are purposefully planned. What is the purpose and what is the plan? One can speculate and assume it is a plan to manipulate weather patterns. One can speculate it is a plan to mark the skies so satellites above the clouds know where to beam down the frequency from the sky to Earth. If it is not weather control or satellite configurations, then do we speculate interplanetary contact discussed in Chapter 11? For example, is there a need to mark the skies to be in contact with an alien presence we cannot see on an everyday level? On a daily basis most people don't look up at the sky anyway because they are stuck looking down at a cell phone.

Whatever the speculation, this is an easy way for a pathogen to spread. Some of the common chemicals analyzed from chemtrail spray are barium and aluminum. An excessive amount of barium is toxic and can affect the nervous and immune systems and the heart. In terms of global surveillance, it has been noted that during the first Gulf War, barium was fed to Iraqi insurgents so they could be tracked by electromagnetic frequency devices. Chemtrails, therefore, can be used as part of the surveillance system (Marrs, 2015).

Along with barium, small nanoparticles of aluminum can enter the nervous system and attribute to a list of neurodegenerative diseases, including Alzheimer's disease, Parkinson's disease and Amyotropic Lateral Sclerosis. Furthermore, why would a company like Monsanto need to develop genetically-modified aluminum-resistant seeds? As suggested by Marrs (2015), it is more than a coincidence and the environment could be poisoned deliberately in order to enable Monsanto to reap greater profits and gain even more control over the world's food supply.

At the level of the United States Federal Government, there will not be any public admission, but the Department of Defense and the United States military have been known to systematically blanket the skies with stratospheric aerosol geoengineering. These are NOT contrails from commercial jet engines. The confusion between contrails and chemtrails has muddle the story to make people think this is a hoax. Even an Ohio representative, Dennis Kucinich, who also ran for President of the United States, had difficulty getting bills passed to ban these exotic weapons. It is too ironic and coincidental for the Department of Defense's 'Vision for 2020' to be written in 2001. Kucinich failed to get these weapons to be exposed to the public as a problem. The failed bill was written back in 2001, and now we have the full blown COVID-19 pandemic to manifesting in 2020.

Periodically, you may have a news report on your local news about some "unexplained" mysterious residue in your community. For example, Marrs (2015) has uncovered anecdotal reports from news channels indicating strange white sticky substances being found on homes, yards, vehicles and shrubbery. Chemical analysis found this substance contained aluminum oxide, barium, polymers, and traces of pathogens. The spray from the chemtrails seemed to make people in the vicinity experience ill effects such as asthma, fatigue, headaches, dizziness, joint pain, and various flu-like symptoms. Many pathogens have been found in the upper atmosphere which is an indication that illnesses brought on by chemtrails could be part of an intentional program.

Since we have the current leadership lying about a wide range of issues, we cannot expect to have anyone from the Federal Government to be focused on the true welfare of its citizenry. If there was care, we would have universal healthcare to prudently and effectively get through this pandemic. In fact, with this theory that the citizenry is being exposed to chemical spraying as a covert means of inoculation, germ warfare experts and medical authorities agree that a high-altitude spraying program is an inefficient method for distributing bacteria and viruses. We may never get to the truth, but the problem has certainly spread world-wide and too fast.

When population control experiments "go wild," a destructive pandemic can be unleashed to change the world. We say silent ethnic weapons because research does indicate that there are ethnic differences in response to drugs and pathogens. Under the guise of these speculations is the reality of population control. There are real-life people like Bill and Melinda Gates who, strangely, "worry" about the stable presence of melanin-dominant people on the planet. There should be no reason for a white billionaire in the computer industry to have an interest in the biology of melanin-dominant people, unless it is a greater agenda to eliminate them. Do your research and read the history for yourself about the Gates agenda. It is too ironic for Bill Gates to resign from the Microsoft Board and the Berkshire Board weeks after the COVID-19 pandemic hit the shores of North America. He is from the State of Washington and that state has had mysterious cases in Senior Assisted Living facilities and other transmissions have occurred with no obvious connection to travelers from Wuhan, China. In fact, the city of Seattle is falling apart on multiple social levels, so Gates should keep his mind, money and focus on his own backyard because Seattle is NOT becoming a place people desire to live.

Overall, melanin-recessive people with money who are billionaires have time to devise plans to control others from staying away from their empire. Culturally speaking, I do not believe that I, as a melanin-dominant billionaire, would desire to control the world. On the other hand, it makes perfect sense for a melanin-recessive person to contemplate the genetic demise of White people on the planet. If this is another speculative theory, then prove Dr. F.C. Welsing wrong in her Cress Theory of Color Confrontation (1991). How much more evidence is needed to validate the xenophobic and fragile nature of a people who feel as though they are being eliminated? Whether a billionaire or poor, subconsciously, melanin-recessive people are fighting for genetic survival. The creation of pandemics demonstrates that survival is by any means necessary.

Just like viruses are real, the need for genetic survival is real and can make people do diabolical things. Current news reports have revealed that white fragile individuals have concocted plans to purposefully contaminate nonwhite communities with COVID-19 by smearing saliva on doorknobs, as an example of devious thinking. In a time of world crisis, it is sad and unfortunate that this mindset cannot be laid to rest. Whether billionaires or poor as dirt, melanin-recessive genetic survival thought patterns are a dysfunctional threat to humanity. The books have been written, the stories have already been told and the movies have been created. It is now time to battle against the compiled sources that have forewarned us that this time would come to rise out of this chaos.

Rise Above the Fear Frequency

Proof that we are all pawns in the game can be related to the unfortunate response to fight over toilet paper in the store during the COVID-19 pandemic. If there was anything to fight over, it should have been the live green and colorful foods in the produce section of the supermarket. Dead meat, cheese products, cow's milk and canned as well as frozen products were sold out, yet fruits and vegetables remained stocked. The sheep mentality to wait for the medical-pharmaceutical-industrial complex to create a vaccine to market is not wise. You have the power to heal by building your defense mechanism system. Brilliant community health activists such as Tamika Moseley (2019) and J.A. Diouck (2018) have health plans to follow. They are alive, living and willing to guide us toward having nature's nutritional beauty to keep us healthy.

More initiative must be taken to get well and stay well and not to be gullible to wait on a treatment from the very entities that caused this problem to flourish in the first place. Older people are more likely to fall in line with a government plan for a mass vaccination program, and it is older people being more impacted by the current COVID-19 virus. Older people are on medications and taking synthetic drugs lowers the body's natural defense system. When taking prescribed drugs, patients are forewarned about the multiple side effects, but what is most glaringly unreported is the lowering of the drug takers immune response. Elderly people taking medications, therefore, will succumb to COVID-19 quicker than younger people.

There has been toxic poisoning on a worldwide scale. The human metabolic system has been bombarded by synthetic drugs, industrial waste, electromagnetic fields generated by power lines, mobile phone towers and other electronic devices (Banks, 2010). Although some doubt the impact, the combination of the 5G technology could be putting unremitting pressure on the human immune system. As a result, vulnerable people are more susceptible to external threats and less responsive to novel biological attacks.

Nothing happens in isolation. Scientific explanations often have multiple causes for events, so there is no coincidence in science. Black people isolated in an Alabama community developed Syphilis in the 1930s and having it went untreated for 40 years. That is no coincidence. To believe that some random green monkey bit a black person in Africa and thereby causing the world to experience an AIDS epidemic is not by chance.

The conditions must be perfect for a disease to flourish. For example, the entire world is impacted by globalization, the massively controlled food industry and deforestation. These seemingly unrelated factors are intricately linked to create an environment for new diseases to form. For example, sick animals raised in close quarters and the destruction of natural habitats can cause more animals to be in close proximity to humans who can now easily travel across borders in a swift manner. These are the types of factors that can create the perfect scenario for a pandemic to spread.

To study all of this, research scientists conduct experiments to see what would happen. The research can be conducted to enhance humanity, or it can be used in a diabolical way to control the human experience. The naïve mind, however, can be easily manipulated to believe any form of propaganda. The years of research before this book, the research in this book, and the research that will come after this book is subject to the interpreter of the information. The mind of the person is critical, and it must be set at a certain frequency for information to enter the brain. In relation to the spread of COVID-19, I can present you with a plethora of overwhelming evidence that this event was planned, but the corporate-controlled media will present innocuous news stories, or a standard counterargument espoused by governmental influence. As a result of a controlled media, it is easy to see how a person could have cognitive blindness and a myopic view of the world. We must fight against this because it can be life-threatening to your existence.

CONCLUSION

By way of conclusion, there really is no end. The only real conclusion will be when we experience another cataclysm which will cleanse the planet or destroy everything in existence. The social structure of our world is so upside down that it is difficult to determine if there will ever be righteousness re-established on planet Earth. Without proper education, the continual attacks on people with pigment will be an ongoing theme. With many elected right-wing governments that have moved into the decade of 2020, the state of justice is dismal. Gentrification and the housing crisis in major metropolitan cities are compounded by the global immigrant crisis throughout the world. Essentially, people are searching for a better place to live and the greedy decisions made by greedy people in large corporations are making it difficult for common people to survive. Until justice is established, we will have continuous conflict between the haves and the have nots.

In fact, we have a historical precedent for the clash of cultures and a contemporary analysis appears that we do not have an "intelligent" resolution. It seems like enemies are perpetual; therefore, we might not be intelligent beings. If the existence of other universes is eventually found, their thinking would not be contaminated by our views. We can only speculate, but perhaps other life forms have learned to work together and not exploit one another.

Although the author was focused on melanin-dominant people as the theme in *Pigment Power*, we are not excluding the universal nature of righteousness. Basically, those who are the fathers and mothers of civilization must return to a state of consciousness to help the children of humanity to see a different view of life. What humans have done; humans can do. Therefore, it is time to stop discrimination and to prepare for the development of a new humanity.

We began with a sociopolitical discussion to set the tone for the controversy with color in our society. Pigment politics will only exist with an uneducated population. All over the world people are yearning for better leadership and a better outlook on life. In this information age, we are on the precipice of killing ourselves with data overload. In some way, we must find a way to tune into self and not get distracted with the bombardment of information.

There has been too much animosity, jealousy, hate and misinformation about people who possess melanin. This book is about showing the positive nature of all elements related to melanin. Whether you "got melanin" or you lack it, the science behind melanin impacts every living entity on the planet. We described how it is in the nervous system and throughout the human body to enhance the human experience. In addition, the presence of the right light as a source of energy can enhance our mental state, and we must continually look for ways to improve life. Without melanin, there is much lack in the human experience on multiple levels of existence. Pigmentation is powerful, so we need to boost our thoughts on how darkness is for protection against the dangerous elements in our environment. The global warming crisis will be a serious wake up call for all, especially those who are melanin-recessive. In the context of the environment, we also provided a substantive review of research on technology that has been influenced by the mechanisms by which natural melanin functions. In Chapter 11, we gave a provocative view on the origins of humanity and the cosmic connections.

If there is a cosmic form of intelligence that will one day return to Earth, it is a high probability that we do not think in the same manner. In fact, we could actually meet the cosmic life on their landscape before we are contacted by them. What would be our interaction with any cosmic life forms from another galaxy? Intelligence is a reified term, so the definition may not due justice. For us to suggest that there are "other" intelligent beings, is to suggest that we are intelligent.

An objective assessment of the human experience would make me doubt our superiority over other life forms. In the so-called most powerful country on the planet, it is unfortunate that we did not elect a righteous citizen to run the operations of this country as the 45th President. Even while being embarrassingly impeached, #45 played games with humanity by creating a new branch of the military called the

Space Force. People are dying and starving in America, and this President desires to spend tax money on a war in the sky.

I seriously doubt this type of intelligence will make for friendly contact with an unknown life form. In recent times, the USA sponsored killing of brown people in other countries (Libya, Iraq and Iran) has lowered our intelligence on what it means to be harmonious. If melanin-recessive warmongers can joyously praise the killing of people who do not look like them, imagine the xenophobic reply to an extraterrestrial? Unfortunately, this diseased state of mind has had a domino effect on Earth. Other countries like Brazil, Philippines, India and Great Britain, for example, have also elected leadership that is arguably setting back social progress that has been made over several decades. Global white fragility is in full effect, and unfortunately, money is the rule of law. Under these circumstances, the haves will continue to exploit the have nots.

The assumption is that the haves are intelligent and that is why they have more than others. In retrospect, are people really comparing human intelligence to apes if we think we evolved from a lower species that does not have our language capabilities? We know there are some genes for a protein like FOXP2 that can trigger us to speak and these genes could be the leap from ape to man – if you believe we evolved from apes. The point to make here is, if we were so intelligent, why do we have so much famine when food is in abundance? Why is there so much miseducation when we all have easy push button access to knowledge and information? Why do we have a spiritual death when we have so many religious doctrines supposedly professing the truth? We cannot be so intelligent and still have so much strife on so many levels. I suspect there is a parallel universe to ours out there in the cosmic cohesion of the universe that displays the beauty of what life can be. Before melanin-recessive war-like invaders from the north came down from the harsh elements of the Ice Age, melanin-dominant civilizations such as Sumer, Nile Valley and the Indus Valley were the living examples of what we could be.

We can speculate on the past and we can speculate on the future. What is for certain, however, is we will not last forever. The creati ve force that brought all into existence can also relieve us of our existence.

183

The many lost civilizations that have been ruined on Earth (and perhaps on other planets like Mars) by the ravages of nature are a continual threat to our existence. I believe one day our galaxy will merge with another galaxy and the universe we know will collapse. Before that major merger in the cosmos, another cataclysmic event will create upheaval on planet Earth. It may not be in our lifetime, but when it does occur, those who survive will be forced to start over again. Some people may be "rememberers of the way," and civilization can be rebooted. Almost by divine prophecy, a great catastrophe did occur when this book was complete, the COVID-19 Pandemic.

In reality, starting over may be the best way to deal with chaos in social structures. A disruptive force leads to change and we must prepare ourselves for one of three scenarios:

1) The "so-called" second coming of the Messiah;
2) The presence of interstellar contact with other cosmic life and our response; and
3) The total destruction of the planet from a major cataclysmic event.

Pigment Power is all about enhancing life, so open your third eye and feel the frequency of this information. Corporations and billionaires do not intend for you to think for yourself and to take their power and money away. They create a frequency of fear to halt your desire to be free and sovereign. As concluded by Marrs (2015), these self-styled globalists believe themselves to be more enlightened, entitled by heritage, and therefore, more worthy than others to rule the world.

Subsequently, the frequency pervading the planet in 2020 is fear. This is dangerous because the fear can enhance death and destruction of the human body. We must reverse this because humans can alter their emotions and physical body through willpower. Whether we say willpower or *Pigment Power*, human DNA is controlled not through biology, but through energetic signals from outside the cell, including a person's negative and positive thoughts (Lipton, 2005).

As we conclude, researchers like David Icke (2002) have warned us for years about the times we are living in. Conspiratorial theorist like

Icke have pushed a message to emphasize what happens when there is control and manipulation of our reality. For the topic of overcoming fear, one must first be aware.

A traumatized mind is a suggestible one. When people feel fear, it generates a vibration in the human body and an energy field around the body. The more humanity feels fear in all its forms, the more energy is given to those in control. A dense energy field has a low vibrational frequency that feeds chaos. When living in disharmony, people move further away from a connection with the one divine spirit that pervades nature. By slowing down and meditating people are moving closer to the divine spirit and infinite wisdom. Limit TV watching because it is feeding fear.

The people we constantly see and acknowledge as elected leaders in the media are existing in survival mode. This is how lower-order animals think, just for the survival instinct. News alert, they are going to die too. Billionaires will die too. They bleed like we bleed, so we must not give them god-like control over our existence. We will overcome this viral threat, and we will do all we can to usher in a new humanity. It is difficult to fight through a people who only function in a survival instinct mode, but we must do so. We must raise the frequency using *Pigment Power*. A higher order of frequency can thwart this pandemic and kill off this coronavirus.

It is understood that many people are not awake, and some people are not going to make it. Currently, we are being programmed away from self-healing. The frequency of fear pervading the planet is a conscious decision to accept it. We must change the language to love and a culture of life. Barack Obama won his presidency on three words – Yes We Can. We can win this current fight on three words - We Will Win.

You can choose your scenario to believe in, but what we do know is that the presence of pigment is a blessing in life, and it will add longevity and quality to the human experience. The human race must embrace one another in order for there to be order. But if perpetual enemies are the cosmic orientation embodied in good and evil, then the battle must be waged, and righteousness will always prevail over wickedness. To spend time thinking about evil is the ultimate in death.

For closing words of wisdom, love thyself and know thyself. If this is the last book I ever write, I wrote it like I will never see you again. I am OK with that because we come into the planet as a creative force to say, "leave the planet better than it was before you came into existence." We must make our reality to be <u>indivisible</u> as one with the creator but <u>dual</u> and distinct as your own unique personality. Yes, you are an INDIVI -DUAL, but we are all ONE with the creative force that brought us into existence

Now that you have been endarkened, live and be wise.

BIOSKETCH

T. Owens Moore, Ph.D.

T. Owens Moore, Ph.D. is trained as a Physiological Psychologist. He is a graduate of Lincoln University in PA, and he has a M.S. and a Ph.D. from Howard University. For over 27 years, he has served as a college professor, a biomedical researcher and an African-centered scholar/activist. He is a distinguished educator who appreciates interdisciplinary studies. He has served as an Assistant Professor at Morehouse College in Atlanta GA and an Associate Professor and Chairperson of the Department of Psychology at Clark Atlanta University. From 2012 to the present, he has earned the status of a Full Professor. During his administrative roles, he has served as the Chair of the Department of Psychology at Fayetteville State University in North Carolina and the Department of Psychology at Clark Atlanta University. His specialty courses are both in the field of Neuroscience and African-Centered Psychology.

Throughout his academic career, he has been actively involved in the Departments of Psychology, Biology and Africana and Women's Studies. In addition, he is a co-founding member of the Neuroscience Institute at Morehouse School of Medicine. He was also an active member of the National Science Foundation funded Center for Behavioral Neuroscience in Atlanta GA. Dr. Moore has received federal and private grant support to conduct research on the effects of hormones and neuropeptides on social behavior in rodents. Through his diversified academic training, he has been able to study learning paradigms that can be utilized to enhance the educational performance of students. He is a 2009 U.S. Fulbright Scholar and he is recognized as a social activist on an international level. He has been on four continents as well as the Caribbean Islands to teach and conduct research. He continues to have an impact on students and intellectuals in many countries, such as Salvador, Brazil where he is recognized as a distinguished scholar/activist. He has written numerous scientific articles, and he has been a popular Blog Talk Radio Host on the Melanology Hour with Dr. Jewel Pookrum. He is the author of several books:

The Science of Melanin: Dispelling the Myths, The Science of Melanin (The Second Edition), Dark Matters-Dark Secrets, Why Darkness Matters: The Power of Melanin in the Brain, Clue Seeker: A Journey Back Through Time to Search for an Identity and The SpiderFly Proverbs: A Practical Guide for Loving Relationships.
(For more information see www.drtmoore.com)

REFERENCES

Adameyko, I., Lallemend, F. F., Aquino, J.B., Pereira, J.A., Topilko, P., Muller, T. and Ernfors, P. (2009). Schwann cell precursors from nerve innervation are a cellular origin of melanocytes in skin. *Cell, 139*(2), 366-379.

Adams, H. (1983a). African Observers of the Universe: The Sirius Question. In I. Van Sertima (ed.), *Blacks in Science: Ancient and Modern: Journal of African Civilizations*, April and November, vol. 5, Nos. 1&2. pp. 27-46. Transaction Publishers: New Brunswick, NJ.

Adams, H. (1983b). New Light on the Dogon and Sirius. In I. Van Sertima (ed.), *Blacks in Science: Ancient and Modern: Journal of African Civilizations*, April and November, vol. 5, Nos. 1&2. pp. 47-49. Transaction Publishers: New Brunswick, NJ.

Afrika, L. (1989). *African Holistic Health*. Adesegun, Johnson, and Koram Publishers: Silver Spring, MD.

Afrika, L. (2000). *Nutricide: The Nutritional Destruction of the Black Race*. A&B Publishers Group: New York, NY.

Afrika, L. (2009). *Melanin: What Makes Black People Black!* Seaburn Publishers: Astoria, NY.

Akbar, N. (1984). *Chains and Images of Psychological Slavery*. Mind Productions: Tallahassee, FL.

Akbar, N. (1994). *Light from Ancient Africa*. Mind Productions: Tallahassee, FL.

Alexander, M. (2012). *The New Jim Crow*. The New Press: New York, NY.

Allen, J., Als, H., Lewis, J. and Litwack, L. (2003). *Without Sanctuary*. Twin Palms Publishers: Santa Fe, NM.

Amen, R.U.N. (1992). *An Afrocentric Guide to a Spiritual Union.* Khamit Corp. Bronx: NY.

Anderson, K.G., Rambaut, A., Lipkin, W.I., Holmes, E.C. and Garry, R.F. (2020). The proximal origin of SARS-CoV-2. Nature Medicine. https://www.nature.com/articles/s41591-020-0820-9

Anderson, S.E. (1995). *The Black Holocaust for Beginners.* Writers and Readers: New York, NY.

Ani, M. (1994). *Yurugu: An African-Centered Critique of European Cultural Thought and Behavior.* African World Press: Trenton, NJ.

Arck, P.C., Overall, R., Spatz, K., Liezman, C., Hanjiska, B., Klapp, B.F., Birch-Machin, M.A. and Peters, E.M. (2006). Towards a "free radical theory of graying": melanocyte apoptosis in the aging human hair follicle is an indicator of oxidative stress induced tissue damage. *FASEB Journal*, July, 20(9), 1567-1569.

Asante, M.K. (1999). *The Painful Demise of Eurocentrism.* African World Press: Trenton, NJ.

Bain, C. (2020). https://www.sbs.com.au/news/ninety-nine-per-cent-of-coronavirus-deaths-in-italy-had-pre-existing-medical-conditions-study-finds

Banks, N.T. (2010). *AIDS, Opium, Diamonds and Empire: The Deadly Virus of International Greed.* iUniverse, Inc: Bloomington, IN.

Barnes, C. (1988). *Melanin: The Chemical Key to Black Greatness.* C.B. Publishers: Houston, TX.

Baruti, M.K.B. (2002). *The Sex Imperative.* Akoben House: Atlanta: Georgia.

Baruti, M.K.B. (2003). *Homosexuality and the Effeminization of Afrikan Males.* Akoben House: Atlanta: Georgia.

Baruti, M.K.B. (2005). *Kebuka: Remembering the Middle Passage through the Eyes of our Ancestors.* Akoben House: Atlanta: Georgia.

Bauval, R. and Brophy, T. (2011). *Black Genesis: The Prehistoric Origins of Ancient Egypt.* Bear and Company: Rochester, VT.

Bellei, B. and Picardo, M. (2019). Premature cell senescence in human skin: dual face chronic acquired pigmentary disorders. *Ageing Research Reviews*, Nov. 13., Epub.

Benedito, E., Jimenez-Cervantes, C., Perez, D., Cubillana, J.D., Solano, F., Jimenez-Cervantes, J., Meyer zum Gottesberge, A.M., Lozano, J.A. and Garcia-Borron, J.C. (1997). Melanin formation in the inner ear is catalyzed by a new tyrosine hydroxylase kinetically and structurally different from tyrosinase. *Biochim Biophys Acta*, Jul. 19, *1336*(1), 59-72.

Ben-Jochannan, Y.A.A. (1989). *Black Man of the Nile and his Family.* Black Classic Press: Silver Spring, MD.

Blackmon, D.A. (2008). *Slavery By Another Name.* Anchor Books: New York, NY.

Bradley, M. (1978). *The Iceman Inheritance: Prehistoric Sources of Western Man's Racism, Sexism and Aggression.* Kayode Publications: New York, NY.

Brainard, G.C., Sherry, D., Skwerer, R.G., Waxler, M., Kelly, K. and Rosenthal, N.E. (1990). Effects of different wavelengths in seasonal affective disorder. *Journal of Affective Disorders*, Dec., *20*(4). 209-216.

Breedlove, S. M. (2017). *Foundations of Neural Development.* Sinauer Associates: Sunderland, MA.

Bunson, M.R. (2002). *Encyclopedia of Ancient Egypt: Revised Edition.* Facts on File, Inc. New York: NY.

Burns, R. (2019). New Military service Space Force launched. *Atlanta Journal-Constitution*, Section A, pg. 3.

Butler, H. (2009). *When Rocks Cry Out*. Stone River Publishing: Fort Worth, TX.

Bynum, E.B. (1999). *The African Unconscious: Roots of Ancient Mysticism and Modern Psychology.* Cosimo Books: New York, NY.

Bynum, E.B., Brown, A., King, R. and Moore. T.O. (2005). *Why Darkness Matters: The Power of Melanin in the Brain*. African American Images: Chicago, IL.

Bynum, E.B. (2012). *Dark Light Consciousness: Melanin, Serpent Power, and the Luminous Matrix of Reality*. Inner Traditions: Rochester, VT.

Cario-Andre, M., Bessou, S., Gontier, E., Maresca, V., Picardo, M. and Taieb, A. (1999). The reconstructed epidermis with melanocytes: a new tool to study pigmentation and photoprotection. *Cell and Molecular Biology*, Nov., *45*(7), 931-942.

Carr, C. (2007). *Our Town*. Three Rivers Press: New York: NY.

Carr, F.W. (2003). Germany's Black Holocaust 1890-1945. Scholar Technological Institute of Research (www.stirinc.org)

Cass, H. and Holford, P. (2002). *Natural Highs*. Avery: New York, NY.

Chae, D.H., Epel, E.S., Nuru-Jeter, A.M., Lincoln, K.D., Taylor, R.J., Lin., J., Blackburn, E.H. and Thomas, S.B. (2016). Discrimination, mental health, and leukocyte telomere length among African American men. *Psychoneuroendocrinology*, Jan., *63*, 10-16.

Chandler, W.B. (1999). *Ancient Future: The Teachings and Prophetic Wisdom of the Seven Hermetic Laws of Ancient Egypt.* Black Classic Press: Baltimore, MD.

Cheng, C., Li, S., Nie, S., Zhao, W., Sun. S. and Zhao, C. (2012). General and biomimetic approach to biopolymer-functionalized graphene oxide nanosheet through adhesive dopamine. *Biomacromolecules*, Dec. 10, *13*(12), 4236-46.

Chu, J. (2019). MIT engineers develop "blackest black" material to date. http://news.mit.edu/2019/blackest-black-material-cnt-0913

Clarke, J.H. (1996). *Critical Lessons in Slavery and the Slave Trade*. Native Sun Publishers: Richmond, VA.

Cone, R.D. (Ed.)(2000). *The Melanocortin Receptors*. Humana Press: Totowa, NJ.

Diez Roux, A.V., Ranjit, N., Jenny, N.S., Shea, S., Cushman, M., Fitzpatrick, A., and Seeman, T. (2009). Race/ethnicity and telomere length in the multi-ethnic study of atherosclerosis. *Aging Cell*, Jun., *8*(3), 251-257.

Diop, C.A. (1974). *The African Origin of Civilization: Myth or Reality*. Lawrence Hill and Co.: Westport, CT. Originally published in French 1955.

Diop. C.A. (1990). *The Cultural Unity of Black Africa*. Third World Press. Chicago: Illinois. Originally published 1959 in French.

Diop, C.A. (1991). *Civilization or Barbarism: An Authentic Anthropology*. Lawrence Hill Books: Brooklyn, NY.

Diouck, J.A.H. (2018). *The Melanin Guide to Spiritual Awakening*. Lulu Press: Morrisville, NC.

Disner, S.G., Beevers, C.G. and Gonzalez-Lima, F. (2016). Transcranial laser stimulation as neuroenhancement for attention bias modification in adults with elevated depression symptoms. *Brain Stimulation*, Sep-Oct, *9*(5), 780-787.

Dray, P. (2003). *At the Hands of Persons Unknown*. The Modern Library: New York, NY.

Dubey, S. and Roulin, A. (2014). Evolutionary and biomedical consequences of internal melanins. *Pigment Cell Melanoma Research*, May, *27*(3), 327-338.

Dudley, D. (1992). *History of the First Council of Nice*. A&B Publishers: Brooklyn, NY. (Originally published 1922, Peter Eckler Publishing Co.)

Duncan, P.D. (1994). *Blacks: The Race from Beyond the Stars*. Harlo Press: Detroit, MI.

Erbele, I.D., Lin, F.R., Agrawal, Y., Francis, H.W., Carey, J.P. and Chien, W.W. (2016). Racial differences of pigmentation in the human vestibular organs. *Otolaryngology Head and Neck Surgery*, Sep., *155*(3), 479-484.

Evans, M.J. (2016). *Zecharia Sitchin and the Extraterrestrial Origins of Humanity*. Bear and Company: Rochester, VT.

Feldman, R.S. and Quenzer, L.F. (1984). *Fundamentals of Neuropsychopharmacology*. Sinauer Associates, Inc: Sunderland, MA.

Feng, L. and Liu, Z. (2011). Graphene in biomedicine: opportunities and challenges. *Nanomedicine,* Feb 6, *6*(2), 317-24.

Finch, C.S. (1991). *Echoes from the Old Darkland*. Khenti, Inc: Decatur, GA.

Finch, C.S. (1998). *The Star of Deep Beginnings: The Genesis of African Science and Technology*. Khenti, Inc: Decatur, GA.

Frenk, E. and Schellhorn, J.P. (1969). Morphology of the epidermal melanin unit. *Dermatologica*, *139*(4), 271-277.

Fuller, N. (1969). *The United Independent Compensatory Code/System/Concept*. Copyrighted, Library of Congress.

Gasque, P. and Jaffar-Bandjee. M.C. (2015). The immunology and inflammatory responses of human melanocytes in infectious diseases. *The Journal of Infection.* Oct;*71*(4):413-21. doi: 10.1016/j.jinf.2015.06.006. Epub 2015 Jun 16.

Gebreab, S.Y., Riestra, P., Gaye, A., Khan, R.J., Xu, R., Davis, A.R., Quarells, R.C., Davis, S.K. and Gibbons, G.H. (2016). Perceived neighborhood problems are associated with shorter telomere length in African American women. *Psychoneuroendocrinology*, Jul., *69*, 90-97.

Ginzburg, R. (1988). *100 Years of Lynchings*. Black Classic Press: Baltimore, MD.

Glimcher, M.E., Kostick, R.M. and Szabo, G. (1973). The epidermal melanocyte system in newborn human skin. A quantitative histologic study. *Journal of Investigative Dermatology*, Dec., *61*(6), 344-347.

Godic, A. Poljsak, B., Adamic, M. and Dahmane, R. (2014). The role of antioxidants in skin cancer prevention and treatment. *Oxid Med Cell Longev*, March 26, Epub.

Gould, E. (2007). How widespread is adult neurogenesis in mammals? *Nature Reviews Neuroscience*, 8(6), 481-488.

Gould, E., Cameron, H.A., Daniels, D.C., Woolley, C.S. and McEwen, B.S. (1992). Adrenal hormones suppress cell division in adult rat dentate gyrus. Journal of Neuroscience, 12(9), 3642-3650.

Griaule, M. and Dieterlen, G. (1986). *The Pale Fox*. The Continuum Foundation: Chino Valley, AZ. Originally published in French 1965.

Guarino, B. (2019). Cave wall may show oldest story ever told. *The Atlanta Journal Constitution*, December 15, A17.

Gunasinghe, R.N., Reuven, D.G., Suggs, K. and Wang, X. (2012). Filled and empty orbital interactions in a planar covalent organic framework on graphene. *Journal of Physical Chemistry Letters,* 3(20), 3048-52.

Hancock, G. (2015). *Magicians of the Gods*. Thomas Dunne Books. New York, NY.

Hawking, S. (2001). *The Universe in a Nutshell*. Bantam Books: New York, NY.

Heater, B. (2017). https://techcrunch.com/2017/12/09/graphene-running-shoes-will-hit-the-market-next-year/

Helmer, A., Slater, N., Smithgall, S. (2018). A Review of ACE Inhibitors and ARBs in Black Patients With Hypertension. *The Annals of Pharmacotherapy.* 2018 Nov;*52*(11):1143-1151. doi: 10.1177/1060028018779082. Epub 2018 May 29.

Henriksen, T.E., Skrede, S., Fasmer, O.B., Schoeyen, H., Leskauskaite, I., Bjorke-Bertheussen, J., Assmus, J., Hamre, B., Gronli, J. and Lund, A. (2016). Blue-blacking glasses as additive treatment for mania: randomized placebo-controlled trial. *Bipolar Disorders Journal*, May, *18*(3), 221-232.

Hill, J. (2019). It's time for black athletes to leave white colleges. *The Atlantic*. October Issue.

Horowitz, L. (1996). *Emerging Viruses: AIDS and EBOLA: Nature, Accident, or Intentional*. Medical Veritas International: Las Vegas, NV.

Houston, D.D. (1985). *Wonderful Ethiopians of the Ancient Cushite Empire*. Black Classic Press: Baltimore, MD.

Huang, L., Xi, Y., Peng, Y., Yang, Y., Huang, X., Fu, Y., Tao, Q., Xiao, J., Yuan, T., An, K., Zhao, H., Pu, M., Xu, F., Xue, T., Luo, M., So, K.F. and Ren, C. (2019). A visual circuit related to habenula underlies the antidepressive effects of light therapy. *Neuron*, Feb. 9, S0896-6273(19), 30064-9, Epub.

Icke, D. (2002). *Alice in Wonderland and the World Trade Center Disaster*. Bridge of Love Publications: Wildwood, MO.

Iniesta, R., Campbell, D., Venturini, C., Faconti, L., Singh, S., Irvin, M.R., Cooper-DeHoff, R.M., Johnson, J.A., Turner, S.T., Arnett, D.K., Weale, M.E., Warren, H., Munroe, P.B., Cruickshank, K., Padmanabhan, S., Lewis, C., and Chowienczyk, P. (2019). Gene Variants at Loci Related to Blood Pressure Account for Variation in Response to Antihypertensive Drugs Between Black and White Individuals. *Hypertension*. Sep;*74*(3):614-622.

Irgens, L.M. (1984). The discovery of *Mycobacterium leprae*: A medical achievement in the light of evolving scientific methods. *American Journal of Dermatopathology*, *6*(4), 337-343.

Iwata, M., Iwata, S., Everett, M.A. and Fuller, B.B. (1990a). Hormonal stimulation of tyrosinase activity in human foreskin organ cultures. *In Vitro Cell Developmental Biology*, Jun., *26*(6), 554-560.

Iwata, M., Corn, T., Iwata, S., Everett, M.A. and Fuller, B.B. (1990b). The relationship between tyrosinase activity and skin color in human foreskins. *Journal of Investigative Dermatology*, Jul, *95*(1), 9-15.

Jaspin, E. (2007). *Buried in the Bitter Waters*. Basic Books: New York, NY.

Johnson, W.R. (1999). Risk factor identification and primary prevention of stroke in African-American populations. In R.F. Gillum, P.B. Gorelick and E.S. Cooper (Eds.). *Stroke in Blacks*, (pp. 118-128). Karger: Basel, Switzerland.

Joly-Tonetti, N., Wibawa, J.I.D., Bell, M. and Tobin, D.J. (2018). An explanation for the mysterious distribution of melanin in human skin: a rare example of asymmetric (melanin) organelle distribution during mitosis of basal layer progenitor keratinocytes. *British Journal of Dermatology*, Nov., *179*(5), 1115-1126.

Jones, D. (1993). *Culture Bandits II: The Annihilation of Afrikan Images*. Hikeka Press: Philadelphia, PA.

Jordan, C.D., Glover, L.M., Gao, Y., Musani, S.K., Mwasongwe, S., Wilson, J.G., Reiner, A., Diez-Rouz, A. and Sims, M. (2019).

Association of psychosocial factors with leukocyte telomere length among African Americans in the Jackson heart study. *Stress Health*, Apr., *35*(2), 138-145.

Joseph, J., Nadeau, D. and Underwood, A. (2002). *The Color Code*. Hyperion: New York, NY.

Joseph, M. (1989). *Animals In Digestive System*. InnerPeace Books: Miami, FL.

Kambon, K.K.K. (1992). *The African Personality in America: An African-centered Framework*. Nubian Nation Publications: Tallahassee, FL.

Kambon, K.K.K. (1998). *African/Black Psychology in the American Context: An African-centered Approach*. Nubian Nation Publications: Tallahassee, FL.

Kim, Y.J., Khetan, A., Chun, S., Viswanathan, V. Whitacre, J.F. and Bettinger (2016). Evidence of Porphyrin-like structures Natural Melanin pigments using electrochemical fingerprinting. https://doi.org/10.1002/adma.201504650

King, R. (1991). *African Origin of Biological Psychiatry*. Seymour-Smith, Inc.: Germantown, TN.

King, R. (1994). *Melanin: A Key to Freedom*. U.B&U.S. Communications Systems: Hampton, VA.

Lam, R.W., Levitt, A.J., Levitan, R.D., Michalak, E.E., Cheung, A.H., Morehouse, R., Ramasubbu, R., Yatham, L.N., and Tam, E.M. (2016). Efficacy of bright light treatment, fluoxetine, and the combination in patients with nonseasonal major depressive disorder: a randomized clinical trial. *JAMA Psychiatry*, Jan., *73*(1), 56-63.

Laurence, R. (1995). *The Book of Enoch: The Prophet*. Translated by Richard Laurence. Wizards Bookshelf: San Diego, CA. (First translation appeared in 1821).

Le Douarin, N.M. (1980). The ontogeny of the neural crest in avian embryo chimaeras. *Nature, 286*(5774), 663-669.

Lieberman, M.L. (1988). *The Sexual Pharmacy: The Complete Guide to Drugs with Sexual Side Effects*. New American Library: New York, NY.

Lin, B.M., Li, W.Q., Curhan, S.G., Stankovic, K.M., Qureshi, A.A. and Curhan, G.C. (2017). Skin pigmentation and risk of hearing loss in women. *American Journal of Epidemiology*, Jul., *186*(1), 1-10.

Lin, F.R., Mass, P., Chien, W., Carey, J.P., Ferrucci, L. and Thorpe, R. (2012). Association of skin color, race/ethnicity, hearing loss among adults in the USA. *Journal of the Association for Research in Otolaryngology*, Feb., *13*(1), 109-117.

Lipton, B. (2005). *The Biology of Belief. Unleashing the Power of Consciousness, Matter and Miracles*. Mountain of Love: Santa Cruz, CA.

Liu, S. Hu, M., Zeng, T.H., Jang, R., Wei, J., Kong, J. and Chen, Y. (2012a). Lateral dimension-dependent antibacterial activity of graphene oxide sheets. *Langmuir*, Aug 21, *28*(33), 12364-72.

Liu, Z., Hu, C., Zhang, W. and Guo, Z. (2012b). Rapid intracellular growth of gold nanostructures assisted by functionalized graphene oxide and its application for surface-enhanced Raman spectroscopy. *Analytical Chemistry*, Dec. 4, *84*(23), 10338-44.

Mackintosh, J.A. (2001). The antimicrobial properties of melanocytes, melanosomes and melanin and the evolution of black skin. *Journal of Theoretical Biology*, July 21, *211*(2), 101-113.

Marrs, J. (2015). *Population Control*. Morrow: New York, NY.

Martinez-Garcia, M. and Montoliu, L. (2013). Albinism in Europe. *Journal of Dermatology*, May, *40*(5), 319-324.

Masuda, M., Yamazaki, K., Matsunaga, T., Kanzaki, J. and Hosoda, Y. (1995). Melanocytes in the dark cell area of human vestibular organs. *Acta Otolaryngol Supplement*, *519*, 152-157.

McColl, S.L. and Veitch, J.A. (2001). Full-spectrum fluorescent lighting: a review of its effects on physiology and health. *Psychological Medicine*, Aug., *31*(6), 949-964.

McFarland, M.J., Taylor, J., McFarland, C.A.S. and Friedman, K.L. (2018). Perceived unfair treatment by police, race, and telomere length: A Nashville community-based sample of black and white men. *Journal of Health and Social Behavior*, Dec., *59*(4), 585-600.

Meningall, D. (2008). *The Melanin Diet*. Wordclay.

Moore, T.O. (1995). *The Science of Melanin: Dispelling the Myths*. Beckham House Publishers: Silver Spring, MD.

Moore, T.O. (2002). *Dark Matters-Dark Secrets*. Zamani Press: Redan, GA.

Moore, T.O. (2004). *The Science of Melanin: The Second Edition*. Zamani Press: Redan, GA.

Moore, T.O. (2011). *The SpiderFly Proverbs: A Practical Guide for Loving Relationships*. Zamani Press: Redan, GA.

Moseley, T. (2019). *Healing From God's Medicine Cabinet*. SSnaturalhealing: Seattle, WA.

Msezane, A.Z., Felfli, Z., Suggs, K., Tesfamichael, A. and Wang, X.Q. (2012). Gold anion catalysis of methane to methanol. 45(3), 127-135.

Mustakeem, S.M. (2016). *Slavery at Sea: Terror, Sex, and Sickness in the Middle Passage*. University of Illinios Press: Chicago, IL.

Nobles (1976). African Philosophy: Assumptions and paradigms for research on black people. (Eds. L. King and V. Dixon). *Fanon Center*

Publication, J. Alfred Canon Research Conference Proceedings: Los Angeles, CA.

Nobles (2006). *Seeking the Sakhu: Foundational Writings for an African Psychology*. Third World Press: Chicago, IL.

Nur, I.M. (2003). *The Meaning of Blackness*. Anu Publishing Company. Lithonia: Georgia.

Obama, B. (2006). *The Audacity of Hope: Thoughts on Reclaiming the American Dream*. Crown Publisher: New York, NY.

Obama, M. (2018). *Becoming*. Crown Publishing Group: New York, NY.

Obenga, T. (2004). *African Philosophy: The Pharaonic Period: 2780-330 BC*. Per Ankh. African Publishing Cooperative. www.perankhbooks.info

O'Connor, T. (2020). Cuba uses 'wonder drug' to fight coronavirus around the world despite U.S. sanctions. https://www.newsweek.com/cuba-drug-fight-coronavirus-us-sanctions-1493872

O'Hare J.P., O'Brien, I.A., Arendt, J., Astley, P., Ratcliffe, W., Andrews, H., Walters, R., and Corrall, R.J. (1986). Does melatonin deficiency cause the enlarged genitalia of the fragile-X syndrome? *Clinical Endocrinology*, Mar., *24*(3), 327-333.

Olmez, E. and Kurcer, Z. (2003). Melatonin attenuates alpha-adrenergic-induced contractions by increasing the release of vasoactive intestinal peptide isolated rat penile bulb. *Urology Research*, Aug., *31*(4), 276-279.

Orive, et al. (2011). "Naked" gold nanoparticles supported on HOPG: melanin functionalization and catalytic activity. *Nanoscale, 3*, 1708-16.

Ott, J. (1975). Let There Be Light. *Cancer Control Journal. Vol. 3* (6) and Vol. 4 (1). Los Angeles, CA.

Pall,, M.L. (2018). Wi-Fi is an important threat to human health. *Environmental Research*, Jul. *164*, 405-416.

Pane, L.M. (2019). Number of U.S. mass killings in 2019 highest since at least 1970s. *Atlanta Journal Constitution*, December 29, A2.

Pappas, S. (2020). Scientists figure out how coronavirus breaks into human cells. https://www.livescience.com/how-coronavirus-infects-cells.html

Paxman, R. and Harris, J. (1982). *A Higher Form of Killing*. Noonday Press: New York, NY.

Perper, R. (2020). Italy, now under lockdown, has been hit hard by the coronavirus outbreak. It also has one of the world's oldest populations with 60% over age 40. https://www.businessinsider.com/italy-coronavirus-old-population-cases-death-rate-2020-3

Perritano, J. (2019). Your Genes: 100 Things You Never Knew. *National Geographic*: Washington, DC.

Poljsak, B. and Dahmane, R. (2012). Free radicals and extrinsic skin aging. *Dermatol Research Practice*, Feb. 29, Epub.

Pookrum, J. (1993). *Vitamins and Minerals from A to Z*. A&B Books Publishers: Brooklyn, NY.

Presse, F., Hervieu, G., Imaki, T., Sawchenko, P.E., Vale, W. and Nahon, J.L. (1992). Rat-melanin concentrating hormone messenger ribonucleic acid expression: marked changes during development and after stress and glucocorticoid stimuli. *Endocrinology*, Sep., *131*(3), 1241-1150.

Ridley, E.J. (1980). *The Neurological Misadventure of Primordial Man: The Solution to the Problems of Africa and the Rest of the World*. The National Newport News Commentator: Newport News, VA.

Rhoden, W. (2006). *Forty Million Dollar Slaves: The Rise, Fall and Redemption of the Black Athlete*. Crown Publishers: New York, NY.

Rogers, J.A. (1965). *The Five Negro Presidents*. Helga M. Rogers C/O Cahill Law Firm: St. Petersburg, FL.

Roth, A.D., Morant, R. and Alberto, P. (1999). High dose etretinate and interferon-alpha: a phase study in squamous cell carcinomas and transitional cell carcinomas. *Acta Oncology, 38*(5), 613-617.

https://www.sbs.com.au/news/ninety-nine-per-cent-of-coronavirus-deaths-in-italy-had-pre-existing-medical-conditions-study-finds

Schirber, M. (2008). New way to kill viruses: Shake them to death. https://www.livescience.com/7472-kill-viruses-shake-death.html

Schrock, J.M., Adler, N.E., Epel, E.S., Nuru-Jeter, A.M., Lin, J., Blackburn, E.H., Taylor, R.J. and Chae, D.H. (2018). Socioeconomic status, financial strain, and leukocyte telomere length in a sample of African American midlife men. *Journal of Racial and Ethnic Disparities*. Jun., *5*(3), 459-467.

Simoes-Costa, M. and Bronner, M.E. (2015). Establishing neural crest identity: A gene regulatory recipe. *Development, 142*(2), 242-257.

Sitchin, Z. (1991). *The 12ᵗʰ Planet: Book I of the Earth Chronicles*. Bear Company: Rochester, VT. Originally published 1976 by Stein and Day.

Somani, S.M. and Romano, J.A. (Eds.) (2001). *Chemical Warfare Agents: Toxicity at Low Levels*. CRC Press: Baco Raton, FL.

Steenhuysen, J. (2008). Scientists create world's blackest black. https://www.reuters.com/article/us-nano-black-idUSN1555030620080115

Stewart, I. (1996). Pyramid power, people power. *Nature*, September 19. *Vol 383*, p. 218.

Stumpf, W.E. (1988). The endocrinology of sunlight and darkness: complimentary roles for Vitamin D and pineal hormones. *Naturwissenschaften, 75*, 247-251.

Stumpf, W.E. and Privette, T.H. (1989). Light, vitamin D and psychiatry: role of 1,25 dihydroxyvitamin D3 (soltriol) in etiology and therapy of seasonal affective disorder and other mental processes. *Psychopharmacology, 97*, 285-294.

Suggs, K., Reuven, D. and Wang, X.Q. (2011). Electronic properties of cycloaddition-functionalized graphene. *Journal of Physical Chemistry, 115*, 8.

Suggs, K., Person, V. and Wang, X.Q. (2011). Band engineering of oxygen doped single-walled carbon nanotubes. *Nanoscale*, Issue 6.

Sun, D.Q., Zhou, X., Lin, F.R., Francis, H.W., Carey, J.P. and Chien, W.W. (2014). Racial difference in cochlear pigmentation is associated with hearing loss risk. *Otology and Neurotology*, Oct. *35*(9), 1509-1514.

Tachibana, M. (1999). Sound needs melanocytes to be heard. *Pigment Cell Research*, Dec., *12*(6), 344-354.

Takahashi, Y., Sipp, D. and Enomoto. H. (2013). Tissue interactions in neural crest cell development and disease. *Science, 341*(6148), 860-863.

Three Initiates (1988). *The Kybalion*. The Yogi Publication Society. Kindle version. (Originally published in 1908).

Torpy, B. (2019). Dekalb see little value in a wealth of dollar stores. *The Atlanta Journal Constitution*, December 29, B9.

Trosper, J. (2015). The exoplanet files: WASP-12. https://futurism.com/the-exoplanet-files-wasp-12b

Van Neste, D. and Tobin, D.J. (2004). Hair cycle and hair pigmentation: dynamic interactions and changes associated with aging. *Micron, 35*(3), 193-200.

Walker, R. (2006). *When We Ruled*. Black Classic Press: Baltimore, MD.

Weiss, R. (2005). Scientist find a DNA change that accounts for white skin. *Washington Post,* December 16.

Welsing, F.C. (1991). *The Isis Papers: The Keys to the Colors*. Third World Press: Chicago, IL.

Whitaker, S and Fleming, J. (2005*). MediSin*. Divine Protection Publications: Wildomar, CA.

Wolf, F.A. (1988). *Parallel Universes*. Simon & Shuster: New York, NY.

Wright, B.E. (1984). *The Psychopathic Racial Personality and Other Essays*. Third World Press: Chicago, Il.

Wright, B. (1990). *Black Robes, White Justice*. Carol Publishing Group: New York, NY.

Yahya, H. (2007). *Atlas of Creation, Vol.1*. Global Publishing: Istanbul, Turkey.

Yang, K., Zhang, S., Zhang, G., Sun, X., Lee, S.T. and Liu, Z. (2010). Graphene in mice: ultrahigh in vivo tumor uptake and efficient photothermal therapy. *Nanoscale Letters*, Sep 8, *10*(9), 3318-23.

Yin, L., Coelho, S.G., Ebsen, D., Smuda, C. Mahns, A. Miller, S.A., Beer, J.Z., Kolbe, L. and Hearing, V.J. (2014). Epidermal gene expression and ethnic pigmentation variations among individuals of Asian, European and African ancestry. *Experimental Dermatology*, Oct., *23*(10), 731-735.

Zhang, Y., Nayak, T.R., Hong, H. and Cai, W. (2012a). Graphene: a versatile nanoplatform for biomedical applications. *Nanoscale,* Jul 7, *4*(13), 3833-42.

Zhang, B. Tang, D., Liu, B., Cui, Y., Chen, H. and Chen G. (2012b). Nanogold-functionalized magnetic beads with redox activity for sensitive electrochemical immunoassay of thyroid-stimulating hormone. *Analytical Chim Acta*, Jan. 20, *711*, 17-23.

Zhou, J., Shang, J., Song, J. and Ping, F. (2013). Interleukin-18 augments growth ability of primary human melanocytes by PTEN inactivation through the AKT/NFkB pathway. *International Journal of Biochemistry and Cell Biology*, Feb. *45*(2), 308-316.

Zhou, J., Ling, J. and Ping., F. (2016). Interferon-gamma attenuates 5-hydroxytrytamine-induced melanogenesis in primary melanocyte. *Biological and Pharmaceutical Bulletin*, *39*(7), 1091-1099.

INDEX

Welsing, Frances Cress, 7, 10, 14, 27, 63, 78-79, 128, 177
West, Kanye, 67
Wetiko Disease, 19
White Dwarf, 125
Williams, Doug, 25, 67
Williams, Serena, 27
Wilson, Russell, 25
Woods, Tiger, 26
World War III, 152
Wright, Jeremiah, 13

Z
Zika virus, 37

www.ingramcontent.com/pod-product-compliance
Lightning Source LLC
Chambersburg PA
CBHW020702270326
41928CB00005B/233